W9-DJI-235

CHARIOTS OF THE GODS?

CHARIOTS OF THE GODS?

Unsolved Mysteries of the Past

ERICH VON DÄNIKEN

Translated by Michael Heron

G. P. Putnam's Sons
New York

FIRST AMERICAN EDITION 1970

First published in Germany by Econ-Verlag under the title
Erinnerungen an die Zukunft

First published in Great Britain
1969 by Souvenir Press Ltd., 95 Mortimer
Street, London, W.1 and simultaneously
in Canada by The Ryerson Press, Toronto 2,
Canada

Fourth Printing

SBN: 399-10128-4

Library of Congress Catalog Card Number: 70-81645

PRINTED IN THE UNITED STATES OF AMERICA

Contents

Illustrations will be found following pages 64 and 128.

Introduction

IT took courage to write this book, and it will take courage to read it. Because its theories and proofs do not fit into the mosaic of traditional archaeology, constructed so laboriously and firmly cemented down, scholars will call it nonsense and put it on the Index of those books which are better left unmentioned. Laymen will withdraw into the snail shell of their familiar world when faced with the probability that finding out about our past will be even more mysterious and adventurous than finding out about the future.

Nevertheless, one thing is certain. There is something inconsistent about our past, that past which lies thousands and millions of years behind us. The past teemed with unknown gods who visited the primeval earth in manned spaceships. Incredible technical achievements existed in the past. There is a mass of know-how which we have only partially rediscovered today.

There is something inconsistent about our archaeology! Because we find electric batteries many thousands of years old. Because we find strange beings in perfect space suits with platinum fasteners. Because we find numbers with fifteen digits—something not registered by any computer. But how did these early men acquire the ability to create them?

There is something inconsistent about our religion. A feature common to every religion is that it promises help and salvation to mankind. The primitive gods gave such promises, too. Why didn't they keep them? Why did they use ultramodern weapons on primitive peoples? And why did they plan to destroy them?

Let us get used to the idea that the world of ideas which has grown up over the millennia is going to collapse. A few years of accurate research has already brought down the mental edifice in which we had made ourselves at home. Knowledge that was hidden in the libraries of secret societies is being rediscovered. The age of space travel is no longer an age of secrets. We have now landed on the moon. Space travel, which aspires to suns and stars, also plumbs the abysses of our past for us. Gods and priests, kings and heroes, emerge from the dark chasms. We must challenge them to deliver up their secrets, for we have the means to find out all about our past, without leaving any gaps, if we really want to.

Modern laboratories must take over the work of archaeological research. Archaeologists must visit the devastated sites of the past with ultrasensitive measuring apparatus. Priests who seek the truth must again begin to doubt everything that is established.

The gods of the dim past have left countless traces which we can read and decipher today for the first time because the problem of space travel, so topical today, was not a problem, but a reality, to the men of thousands of years ago. I claim that our forefathers received visits from the universe in the remote past, even though I do not yet know who these extraterrestrial intelligences were or from which planet they came. I nevertheless proclaim that these "strangers" annihilated part of mankind existing at the time and produced a new, perhaps the first, *homo sapiens*.

This assertion is revolutionary. It shatters the base on which a mental edifice that seemed to be so perfect was constructed. It is my aim to try to provide proof of this assertion.

My book would not have been written without the encouragement and collaboration of many people. I should like to thank my wife, who has seen little of me at home during the last few years, for her understanding. I should like to thank my friend Hans Neuner, my traveling companion for many thousands of miles, for his unfailing and valuable help. I should like to thank Dr. Stehlin and Louis Emrich for their continuous support. I should like to thank all the NASA personnel at Houston, Cape Kennedy, and Huntsville who showed me around their magnificent scientific and technical research centers. I should like to thank Wernher von Braun, Willy Ley, and Bert Slattery. I should like to thank all the countless men and women around the globe whose practical help, encouragement, and conversation made this book possible.

ERICH VON DÄNIKEN

CHARIOTS OF THE GODS?

I

Are There Intelligent Beings in the Cosmos?

IS it conceivable that we world citizens of the twentieth century are not the only living beings of our kind in the cosmos? Because no homunculus from another planet is on display in a museum for us to visit, the answer, "Our earth is the only planet with human beings," still seems to be legitimate and convincing. But the forest of question marks grows and grows as soon as we make a careful study of the facts resulting from the latest discoveries and research work.

On a clear night the naked eye can see about 4,500 stars, so the astronomers say. The telescope of even a small observatory makes nearly 2,000,000 stars visible, and a modern reflecting telescope brings the light from thousands of millions more to the viewer—specks of light in the Milky Way. But in the colossal dimensions of the cosmos our stellar system is only a tiny part of an incomparably larger stellar system—of a cluster of Milky Ways, one might say, containing some twenty galaxies within a radius of 1,500,000 light-years (1 light-year = the distance traveled by light in a year, *i.e.*, 186,000×60×60×24×365 miles). And even this vast number of stars is small in comparison with the many thousands of spiral nebulae disclosed by the electronic telescope. Disclosed to the present day, I should emphasize, for research of this kind is only just beginning.

Astronomer Harlow Shapley estimates that there are some 10^{20} stars within the range of our telescopes. When Shapley associates a planetary system with only one in a thousand stars, we may assume that it is a very cautious estimate. If we continue to speculate on the basis of this estimate and suspect the necessary conditions for life on only one star in a thousand, this calculation still gives a figure of 10^{14}. Shapley asks: How many stars in this truly "astronomical" figure have an atmosphere suitable for life? One in a thousand? That would still leave the incredible figure of 10^{11} stars with the prerequisites for life. Even if we assume that only every thousandth planet out of this figure has produced life, there are still 100,000,000 planets on which we can speculate that life exists. This calculation is based on telescopes using the techniques available today, but we must not forget that these are constantly being improved.

If we follow the hypothesis of biochemist Dr. Stanley Miller, life and the conditions essential for life may have developed more quickly on some of these planets than on earth. If we accept this daring assumption, civilizations more advanced than our own could have developed on 100,000 planets.

The late Willy Ley, well-known scientific writer, and friend of Wernher von Braun, told me in New York: "The estimated number of stars in our Milky Way alone amounts to 30 billion. The assumption that our Milky Way contains at least 16 billion planetary systems is considered admissible by present-day astronomers. If we now try to reduce the figures in question as much as possible and assume that the distances between planetary systems are so regulated that only in one case in a hundred does a planet orbit in the ecosphere of its own sun, that still leaves 180 million planets capable of supporting life. If we further assume that only one planet in a hundred that might support life actually does so, we should still have the figure of 1.8 million planets with life. Let us further suppose that out of every hundred planets with life there is one on which creatures with the same level of intelligence as *homo sapiens* live. Then even this last supposition gives our Milky Way the vast number of 18,000 inhabited planets."

Since the latest counts give 100 billion fixed stars in our Milky Way, probability indicates an incomparably higher figure than Dr. Ley puts forward in his cautious calculation.

Without quoting fantastic figures or taking unknown galaxies into account, we may surmise that there are 18,000 planets comparatively close to the earth with conditions essential to life similar to those of our own planet. Yet we can go even further and speculate that if only 1 percent of these 18,000 planets were actually inhabited, there would still be 180 left!

There is no doubt about the existence of planets similar to the earth—with a similar mixture of atmospheric gases, similar gravity, similar flora, and possibly even similar fauna. But is it even essential for the planets that support life to have conditions similar to the earth's?

The idea that life can flourish only under terrestrial conditions has been made obsolete by research. It is a mistake to believe that life cannot exist without water and oxygen. Even on our own earth there are forms of life that need no oxygen. They are called anaerobic bacteria. A given amount of oxygen acts like poison on them. Why should there not be higher forms of life that do not need oxygen?

Under the pressure of the new knowledge that is being acquired every day, we shall have to bring our mental world picture up to date. Scientific investigation, concentrated on our earth until very recently, has praised this world of ours as the ideal planet. It is not too hot and not too cold; it has plenty of water; there are unlimited quantities of oxygen; organic processes constantly rejuvenate nature.

In fact, the assumption that life can exist and develop only on a planet like the earth is untenable. It is estimated that 2,000,000 different species of living creatures live on the earth. Of these—this again is an estimate—1,200,000 are "known" scientifically. And among these forms of life known to science there are still a few thousand that ought not to be able to live at all according to current ideas! The premises for life must be thought out and tested anew.

For example, one would think that highly radioactive water would be free from germs. But there are actually some kinds of bacteria which can adapt themselves to the lethal water that surrounds nuclear reactors. An experiment carried out by biologist Dr. Sanford Siegel sounds eery. He re-created the atmospheric conditions of Jupiter in his laboratory and bred bacteria and mites in this atmosphere, which shares none of the prerequisites we have hitherto laid down for "life." Ammonia, methane, and hydrogen did not kill them. The experiments by Dr. Howard Hinton and Dr. Blum, Bristol University entomologists, had equally startling results. The two scientists dried a species of midge for several hours at a temperature of 100° C. Immediately afterward they immersed their "guinea pigs" in liquid helium, which, as is well known, is as cold as space. After heavy irradiation they returned the midges to their normal living conditions. The insects continued their biological vital functions and produced perfectly healthy midges. We also know of bacteria that live in volcanoes, of others that eat stone, and some that produce iron. The forest of question marks grows.

Experiments are going on at many research centers. New proofs that life is by no means bound to the prerequisites for life on our planet are constantly accumulating. For centuries the world appeared to revolve around the laws and conditions that govern life on earth. This conviction distorted and blurred our way of looking at things; it put blinkers on scientific investigators, who unhesitatingly accepted our standards and systems of thought when viewing the universe. Teilhard de Chardin, the epoch-making thinker, suggested that only the fantastic has a chance of being real in the cosmos!

If our way of thinking worked the other way around, it would mean that intelligences on another planet took *their* living conditions as a criterion. If they lived at temperatures of minus 150–200° C, they would think that those temperatures, which are destructive to life as we know it, were essential for life on other planets. That would match up to the logic with which we are trying to illuminate the darkness of our past.

We owe it to our self-respect to be rational and objective. At some time or other every daring theory seemed to be a Utopia. How many Utopias have long since become everyday realities! Of course the examples given here are meant to point out the most farfetched possibilities. Yet once the improbable things that we cannot even conceive of today are shown to be true, as they will be, barriers will fall, allowing free access to the impossibilities the cosmos still conceals. Future generations will find all kinds of life that have never been dreamed of in the universe. Even if we are not there to see it, they will have to accept the fact that they are not the only, and certainly not the oldest, intelligences in the cosmos.

The universe is estimated to be between eight and twelve billion years old. Meteorites bring traces of organic matter under our microscopes. Bacteria millions of years old awake to new life. Floating spores traverse the universe and at some time or other are captured by the gravitational field of a planet. New life has gone on developing in the perpetual cycle of creation for millions of years. Innumerable careful examinations of all kinds of stones in all parts of the world prove that the earth's crust was formed about 4,000,000,000 years ago. Yes, and all that science knows is that something like man existed 1,000,000 years ago! And out of that gigantic river of time it has managed to dam up only a tiny rivulet of 7,000 years of human history, at the cost of a lot of hard work, many adventures and a great deal of curiosity. But what are 7,000 years of human history compared with thousands of millions of years of the history of the universe?

We—the paragons of creation?—took 400,000 years to reach our present state and our present stature. Who can produce concrete proof to show why another planet should not have provided more favorable conditions for the development of other or similar intelligences? Is there any reason why we may not have "competitors" on another planet who are equal or superior to us? Are we entitled to discard this possibility? We have done so up to the present.

How often the pillars of our wisdom have crumbled into dust! Hundreds and hundreds of generations thought that the earth was flat. The iron law that the sun went around the earth held good for thousands of years. We are still convinced that our earth is the center of everything, although it has been proved that the earth is an ordinary star of insignificant size, 30,000 light-years from the center of the Milky Way.

The time has come for us to admit our insignificance by making discoveries in the infinite unexplored cosmos. Only then shall we realize that we are nothing but ants in the vast state of the universe. And yet our future and our opportunities lie in the universe, where the gods promised they would.

Not until we have taken a look into the future shall we be strong and bold enough to investigate our past honestly and impartially.

2

When Our Spaceship Landed on Earth . . .

JULES VERNE, the grandfather of all science-fiction novelists, has become an accepted writer. His fantasies are no longer science fiction, and the astronauts of our day travel around the world in 86 minutes, not 80 days. We are now going to describe what might happen on an imaginary journey by spaceship, yet this imaginary journey will become possible in fewer decades than the time it took to contract Jules Verne's crazy notion of a journey around the world in 80 days to a lightning journey of 86 minutes. But let us not think in terms of such short periods of time! Let us assume that our spaceship will leave the earth for an unknown distant sun in 150 years' time.

The spaceship would be as big as a present-day ocean liner and would therefore have a launching weight of about 100,000 tons with a fuel load of 99,800 tons, *i.e,* an effective payload of 200 tons.

Impossible?

Already we could assemble a spaceship piece by piece while in orbit around a planet. Yet even this assembly work will become unnecessary in less than two decades, because it will be possible to prepare the giant spaceship for launching on the moon. Besides, the basic research for the rocket propulsion of tomorrow is in full swing. Future rocket motors will mainly be powered by nuclear fusion and travel at nearly the speed of light. A bold new method, the feasibility of which has already been shown by physical experiments on individual elementary particles, will be the photon rocket. The fuel carried on board the photon rocket enables the rocket's velocity to approach so close to the speed of light that the effects of relativity, especially the variation in time between launching site and spaceship, can operate to the full. The fuel supplies will be transformed into electromagnetic radiation and ejected as a clustered propulsive jet with the speed of light. Theroetically a spaceship equipped with photon propulsion can reach 99 percent of the speed of light. At this speed the boundaries of our solar system would be blasted open!

An idea that really makes the mind reel. But we who are on the threshold of a new age should remember that the giant strides in technology which our grandfathers experienced were just as staggering in their day: the railways, electricity, telegraphy, the first car, the first airplane. We ourselves heard music in the air for the first time; we see color TV; we saw the first launching of spaceships, and American astronauts actually walking on the moon; and we get news and pictures from satellites that revolve around the earth. Our children's children will go on interstellar journeys and carry out cosmic research in the universities.

Let us follow the journey of our imaginary spaceship, whose goal is a distant fixed star. It would certainly be amusing to try to imagine what the crew of the spaceship did to kill time on their journey. Because however vast the distances they covered and however slowly time might crawl along for those left behind on earth, Einstein's theory of relativity still holds good. It may sound incredible, but time on board the spaceship traveling barely below the speed of light actually passes more slowly than on the earth.

For example, only 10 years pass for our crew on their flight in the universe, whereas 108 years go by for those who stay at home. This shift in time between the space travelers and the people on earth can be calculated by the basic rocket equation worked out by Professor Ackeret:

$$v/w \frac{1 - (1 - t)^{2w/c}}{wc \cdot 11 + (1 - t)^{2w/c} 1}$$

(v = velocity, w = speed of jet, c = speed of light, t = fuel load at launching)

At the moment when our spaceship is approaching the star which is its target, the crew will undoubtedly examine planets, fix their position, undertake spectral analyses, measure forces of gravity, and calculate orbits. Lastly they will choose as landing place the planet whose conditions come closest to those of our earth. If our spaceship consists solely of its payload after a journey of, shall we say, 80 light-years, because all the energy supplies have been used up, the crew will have to replenish the tanks of their spacecraft with fissionable material at their goal.

Let us assume, then, that the planet chosen to land on is similar to the earth. I have already said that this assumption is by no means impossible. Let us also venture the supposition that the civilization of the planet visited is in about the same state of development as the earth was 8,000 years ago. Of course, this would all have been confirmed by the instruments on board the spaceship long before the landing. Naturally our space travelers have also picked on a landing site that lies close to a supply of fissionable matter. Their instruments show quickly and reliably in which mountain ranges uranium can be found.

The landing is carried out according to plan.

Our space travelers see beings making stone tools; they see them hunting and killing game by throwing spears; flocks of sheep and goats are grazing on the steppe; primitive potters are making simple household utensils. A strange sight to greet our astronauts!

But what do the primitive beings on this planet think about the monstrosity that has just landed there and the figures that climbed out of it? Let us not forget that we too were semisavages 8,000 years ago. So it is not surprising when the semisavages who experience this event bury their faces in the ground and dare not raise their eyes. Until this day they have worshiped the sun and the moon. And now something earth-shaking has happened: the gods have come down from heaven!

From a safe hiding place the inhabitants of the planet watch our space travelers, who wear strange hats with rods on their heads (helmets with antennae); they are amazed when the night is made bright as day (searchlights); they are terrified when the strangers rise effortlessly into the air (rocket belts); they bury their heads in the ground again when weird unknown "animals" soar in the air, droning, buzzing, and snorting (helicopters, all-purpose vehicles), and lastly they take flight to the safe refuge of their caves when a frightening boom and rumble resounds from the mountains (a trial explosion). Undoubtedly our astronauts must seem like almighty gods to these primitive people!

Day by day the space travelers continue their laborious work, and after some time a delegation of priests or medicine men will probably approach the astronauts in order to make contact with the gods. They bring gifts to pay homage to their guests. It is conceivable that our spacemen will rapidly learn the language of the inhabitants with the help of a computer and can thank them for the courtesy shown. Yet although they can explain to the savages in their own language that no gods have landed, that no higher beings worthy of adoration have paid a visit, it has no effect. Our primitive friends simply do not believe it. The space travelers came from other stars; they obviously have tremendous power and the ability to perform miracles. They must be gods! There is also no point in the spacemen's trying to explain any help they may offer. It is all far beyond the comprehension of these people who have been so terrifyingly invaded.

Although it is impossible to imagine all the things that might take place from the day of landing onward, the following points might well figure on a preconceived plan:

Part of the population would be won over and trained to help search a crater formed by an explosion for fissionable matter needed for the return to earth.

The most intelligent of the inhabitants would be elected "king." As a visible sign of his power, he would be given a radio set through which he could contact and address the "gods" at any time.

Our astronauts would try to teach the natives the simplest forms of civilization and some moral concepts, in order to make the development of a social order possible. A few specially selected women would be fertilized by the astronauts. Thus a new race would arise that skipped a stage in natural evolution.

We know from our own development how long it would take before this new race became space experts. Consequently, before the astronauts began their return flight to earth, they would leave behind clear and visible signs which only a highly technical, mathematically based society would be able to understand much, much later.

Any attempt to warn our protégés of dangers in store would have little chance of success. Even if we showed them the most horrifying films of terrestrial wars and atomic explosions, it would not prevent the beings living on this planet from committing the same follies any more than it now stops (almost) the whole of sentient humanity from constantly playing with the burning flame of war.

While our spaceship disappears again into the mists of the universe, our friends will talk about the miracle—"The gods were here!" They will translate it into their simple language and turn it into a saga to be handed down to their sons and daughters. They will turn the presents and implements and everything that the space travelers left behind into holy relics.

If our friends have mastered writing, they may make a record of what happened: uncanny, weird, miraculous. Then their texts will relate—and drawings will show—that gods in golden clothes were there in a flying boat that landed with a tremendous din. They will write about chariots which the gods drove over land and sea, and of terrifying weapons that were like lightning, and they will recount that the gods promised to return.

They will hammer and chisel in the rock pictures of what they had once seen: shapeless giants with helmets and rods on their heads, carrying boxes in front of their chests; balls on which indefinable beings sit and ride through the air; staves from which rays are shot out as if from a sun; strange shapes, resembling giant insects, which were vehicles of some sort.

There are no limits to the fantasy of the illustrations that result from the visit of our spaceship. We shall see later what traces the "gods" who visited the earth in our remote antiquity engraved on the tablets of the past.

It is quite easy to sketch the subsequent development of the planet that our spaceship visited. The inhabitants have learned a lot by watching the "gods" surreptitiously; the place on which the spaceship stood will be declared holy ground, a place of pilgrimage, where the heroic deeds of the gods will be praised in song. Pyramids and temples will be built on it—in accordance with astronomic laws, of course. The population increases, wars devastate the place of the gods, and then come generations who rediscover and excavate the holy places and try to interpret the signs.

This is the stage we have reached. Now that we have landed men on the moon, we can open our minds to space travel. We know the effect the sudden arrival of a large ocean-going sailing vessel had on primitive people in, for example, the South Sea Islands. We know the devastating effect a man like Cortés, from another civilization, had on South America. So then we can appreciate, if only dimly, the fantastic impact the arrival of spacecraft would have made in prehistoric times.

We must now take another look at the forest of question marks—the array of unexplained mysteries. Do they make sense as the remains of prehistoric space travelers? Do they lead us into our past and yet link up with our plans for the future?

3

The Improbable World of the Unexplained

OUR historical past is pieced together from indirect knowledge. Excavations, old texts, cave drawings, legends, and so forth were used to construct a working hypothesis. From all this material an impressive and interesting mosaic was made, but it was the product of a preconceived pattern of thought into which the parts could always be fitted, though often with cement that was all too visible. An event must have happened in such and such a way. In that way and no other. And lo and behold—if that's what the scholars really want—it did happen in that way. We are entitled, indeed we ought, to doubt every accepted pattern of thought or working hypothesis, for if existing ideas are not called in question, research is at an end. So our historical past is only relatively true. If new aspects of it turn up, the old working hypothesis, however familiar it may have become, must be replaced by a new one. It seems the moment has come to introduce a new working hypothesis and place it at the very center of our research into the past.

New knowledge about the solar system and the universe, about macrocosm and microcosm, tremendous advances in technology and medicine, in biology and geology, the beginning of space travel—these and many other things have completely altered our world picture in fewer than fifty years.

Today we know that it is possible to make space suits that can withstand extremes of heat and cold. Today we know that space travel is no longer a Utopian idea. We are familiar with the miracle of color television, just as we can measure the speed of light and calculate the consequences of the theory of relativity.

Our world picture, which is already almost frozen into immobility, begins to thaw. New working hypotheses need new criteria. For example, in the future, archaeology can no longer be simply a matter of excavation. The mere collection and classification of finds is no longer adequate. Other branches of science will have to be consulted and made use of if a reliable picture of our past is to be drawn.

So let us enter the new world of the improbable with an open mind and bursting with curiosity! Let us try to take possession of the inheritance the "gods" have bequeathed to us.

At the beginning of the eighteenth century ancient maps which had belonged to an officer in the Turkish navy, Admiral Piri Reis, were found in the Topkapi Palace. Two atlases preserved in the Berlin State Library which contain exact reproductions of the Mediterranean and the region around the Dead Sea also came from Piri Reis.

All these maps were handed over to American cartographer Arlington H. Mallerey for examination. Mallerey confirmed the remarkable fact that all the geographical data were present but not drawn in the right places. He sought the help of Mr. Walters, cartographer in the U.S. Navy Hydrographic Bureau. Mallerey and Walters constructed a grid and transferred the maps to a modern globe. They made a sensational discovery. The maps were absolutely accurate—and not only as regards the Mediterranean and the Dead Sea. The coasts of North and South America and even the contours of the Antarctic were also precisely delineated on Piri Reis' maps. The maps not only reproduced the outlines of the continents but also showed the topography of the interiors! Mountain ranges, mountain peaks, islands, rivers, and plateaus were drawn in with extreme accuracy.

In 1957, the Geophysical Year, the maps were handed over to Jesuit Father Lineham, who is both director of the Weston Observatory and a cartographer in the U.S. Navy. After scrupulous tests Father Lineham, too, could but confirm that the maps were fantastically accurate—even about regions which we have scarcely explored today. What is more, the mountain ranges in the Antarctic, which already figure on Reis' maps, were not discovered until 1952. They have been covered in ice for hundreds of years, and our present-day maps have been drawn with the aid of echo-sounding apparatus.

The latest studies of Professor Charles H. Hapgood and mathematician Richard W. Strachan give us some more shattering information. Comparison with modern photographs of our globe taken from satellites showed that the originals of Piri Reis' maps must have been aerial photographs taken from a very great height. How can that be explained?

A spaceship hovers high above Cairo and points its camera straight downward. When the film is developed, the following picture would emerge: everything that is in a radius of about 5,000 miles of Cairo is reproduced correctly, because it lies directly below the lens. But the countries and continents become increasingy distorted the farther we move our eyes from the center of the picture.

Why is this?

Owing to the spherical shape of the earth, the continents away from the center "sink downward." South America, for example, appears strangely distorted lengthways, exactly as it does on the Piri Reis maps! And exactly as it does on the photographs taken from the American lunar probes.

There are one or two questions that can be answered

quickly. Unquestionably our forefathers did not draw these maps. Yet there is no doubt that the maps must have been made with the most modern technical aid—from the air.

How are we to explain that? Should we be satisfied with the legend that a god gave them to a high priest? Or should we simply take no notice of them and pooh-pooh the "miracle" because the maps do not fit into our mental world picture? Or should we boldly stir up a wasps' nest and claim that this cartography of our globe was carried out from a high-flying aircraft or from a spaceship?

Admittedly the Turkish admiral's maps are not originals. They are copies of copies of copies. Yet even if the maps dated from only the eighteenth century, when they were found, these facts are just as unexplainable. Whoever made them must have been able to fly and also to take photographs!

Not far from the sea, in the Peruvian spurs of the Andes, lies the ancient city of Nazca. The Palpa Valley contains a strip of level ground some 37 miles long and 1 mile wide that is scattered with bits of stone resembling pieces of rusty iron. The inhabitants call this region pampa, although any vegetation is out of the question there. If you fly over this territory—the plain of Nazca—you can make out gigantic lines, laid out geometrically, some of which run parallel to each other, while others intersect or are surrounded by large trapezoidal areas.

The archaeologists say that they are Inca roads.

A preposterous idea! Of what use to the Incas were roads that ran parallel to each other? That intersected? That were laid out in a plain and came to a sudden end?

Naturally typical Nazca pottery and ceramics are found here, too. But it is surely oversimplifying things to attribute the geometrically arranged lines to the Nazca culture for that reason alone.

No serious excavations were carried out in this area until

1952. There is no established chronology for all the things that were found. Only now have the lines and geometrical figures been measured. The results clearly confirm the hypothesis that the lines were laid out according to astronomical plans. Professor Alden Mason, a specialist in Peruvian antiquities, suspects signs of a kind of religion in the alignments, and perhaps a calendar as well.

Seen from the air, the clear-cut impression that the 37-mile-long plain of Nazca made on *me* was that of an airfield!

What is so farfetched about the idea?

Research (= knowledge) does not become possible until the thing that is to be investigated has actually been found! Once it is found, it is tirelessly polished and trimmed until it has become a stone that—miraculously enough—fits exactly into the existing mosaic. Classical archaeology does not admit that the pre-Inca peoples could have had a perfect surveying technique. And the theory that aircraft could have existed in antiquity is sheer humbug to them.

In that case, what purpose did the lines at Nazca serve? According to my way of thinking, they could have been laid out on their gigantic scale by working from a model and using a system of coordinates, or they could also have been built according to instructions from an aircraft. It is not yet possible to say with certainty whether the plain of Nazca was ever an airfield. If iron was used it will certainly not be found, because there is no prehistoric iron. Metals corrode in a few years; stone never corrodes. What is wrong with the idea that the lines were laid out to say to the "gods": "Land here! Everything has been prepared as you ordered"? The builders of the geometrical figures may have had no idea what they were doing. But perhaps they knew perfectly well what the "gods" needed in order to land.

Enormous drawings that were undoubtedly meant as signals for a being floating in the air are found on mountainsides in many parts of Peru. What other purpose could they have served?

One of the most peculiar drawings is carved on the high red wall of the cliffs in the Bay of Pisco. If you arrive by sea, you can make out a figure nearly 820 feet high from a distance of more than 12 miles. If you play at "It looks like . . . ," your immediate reaction is that this sculptor's work looks like a gigantic trident or a colossal three-branched candlestick. And a long rope was found on the central column of this stone sign. Did it serve as a pendulum in the past?

To be honest, we must admit that we are groping in the dark when we try to explain it. It cannot be meaningfully included in existing dogmas, which does not mean to say that there may not be some trick by which scholars could conjure this phenomenon too into the great mosaic of accepted archaeological thinking.

But what can have induced the pre-Inca peoples to build the fantastic lines, the landing strips, at Nazca? What madness could have driven them to create the 820-foot-high stone signs on the red cliffs south of Lima?

These tasks would have taken decades without modern machines and appliances. Their whole activity would have been senseless if the end product of their efforts had not been meant as signs to beings approaching them from great heights. The stimulating question still has to be answered: Why did they do all this if they had no idea that flying beings actually existed?

The identification of finds can no longer remain a matter for archaeology alone. A council of scientists from different fields of research would certainly bring us close to the solution of the puzzle. Exchange of opinions and dialogue would definitely produce illuminating insights. Because scientists do not take the posing of such questions seriously, there is a danger that research will come to no definite conclusions. Space travelers in the gray mists of time? An inadmissible question to academic scientists. Anyone who asks questions like that ought to see a psychiatrist.

But the questions are there, and questions, thank heavens, have the impertinent quality of hovering in the air until they are answered. And there are many inadmissible questions like that. For example, what would people say if there were a calendar which gave the equinoxes, the astronomical seasons, the positions of the moon for every hour and also the movements of the moon, even taking the rotation of the earth into account?

That is no mere hypothetical question. This calendar exists. It was found in the dry mud at Tiahuanaco. It is a disconcerting find. It yields irrefutable facts and proves—can our self-assurance admit such a proof?—that the beings who produced, devised, and used the calendar had a higher culture than ours.

Another quite fantastic discovery was the Great Idol. This single block of red sandstone is longer than 24 feet and weighs 20 tons. It was found in the Old Temple. Again we have a contradiction between the superb quality and precision of the hundreds of symbols all over the idol and the primitive technique used for the building housing it. Indeed it is called the Old Temple because of the primitive technique.

H. S. Bellamy and P. Allan have given a closely reasoned interpretation of the symbols in their book *The Great Idol of Tiahuanaco*. They conclude that the symbols record an enormous body of astronomical knowledge and are based, as a matter of fact, on a round earth.

They conclude that the record fits perfectly Hoerbiger's *Theory of Satellites*, published in 1927, five years before the idol was discovered. This theory postulates that a satellite was captured by the earth. As it was pulled toward the earth it slowed down the speed of the earth's revolutions. It finally disintegrated and was replaced by the moon.

The symbols on the idol exactly record the astronomical phenomena which would accompany this theory at a time when the satellite was making 425 revolutions around the earth in a year of 288 days. They are forced to conclude that the idol records the state of the heavens 27,000 years ago. They write, "Generally, the idol inscriptions give the impression . . . that it had been devised also as a record for future generations."

Here indeed is an object of great antiquity which demands a better explanation than "an ancient god." If this interpretation of the symbols can be substantiated, we must ask: Was the astronomical knowledge really amassed by people who still had a great deal to learn about building, or did it come from extraterrestrial sources? In either case the existence of such a sophisticated body of knowledge 27,000 years ago, demonstrated on both the idol and the calendar, is a staggering thought.

The city of Tiahuanaco teems with secrets. The city lies at a height of more than 13,000 feet, and it is miles from anywhere. Starting from Cuzco, Peru, you reach the city and the excavation sites after several days' travel by rail and boat. The plateau looks like the landscape of an unknown planet. Manual labor is torture for anyone who is not a native. The atmospheric pressure is about half as low as it is at sea level and the oxygen content of the air is correspondingly small. And yet an enormous city stood on this plateau.

There are no authentic traditions about Tiahuanaco. Perhaps we should be glad that in this case acceptable answers cannot be reached by leaning on the crutch of hereditary orthodox learning. Over the ruins, which are incredibly old (how old we do not yet know), lies the mist of the past, ignorance and mystery.

Blocks of sandstone weighing 100 tons are topped with other 60-ton blocks for walls. Smooth surfaces with extremely accurate chamfers join enormous squared stones which are held together with copper clamps. In addition, all the stonework is exceptionally neatly executed. Holes 8 feet long, whose purpose has not been explained thus far, are found in blocks weighing 10 tons. Nor do the 16½-foot-long, worn-down flagstones cut out of one piece contribute to the solution of the mystery that Tiahuanaco conceals. Stone water conduits, 6 feet long and 1½ feet wide, are found scattered about on the ground like toys, obviously by a catastrophe of tremendous dimensions. These finds stagger us by their accurate workmanship. Had our forefathers at Tiahuanaco nothing better to do than spend years—without tools—fashioning water conduits of such precision that our modern concrete conduits seem the work of mere bunglers in comparison?

In a courtyard which has now been restored there is a jumble of stone heads which, on closer observation, appears to be made up of the most varied races, for some of the faces have narrow lips, and some swollen; some long noses, and some hooked; some delicate ears, and some thick; some soft features, and some angular. And some of the heads wear strange helmets. Are all these unfamiliar figures trying to convey a message that we cannot or will not understand, inhibited as we are by stubbornness and prejudice?

One of the great archaeological wonders of South America is the monolithic Gate of the Sun at Tiahuanaco—a gigantic sculpture, nearly 10 feet high and 16½ feet wide, carved out of a single block. The weight of this piece of masonry is estimated at more than 10 tons. Forty-eight square figures in three rows flank a being who represents a flying god.

What does legend say about the mysterious city of Tiahuanaco?

It tells of a golden spaceship that came from the stars; in it came a woman, whose name was Oryana, to fulfill the task of becoming the Great Mother of the earth. Oryana had only four fingers, which were webbed. Great Mother Oryana gave birth to 70 earth children, then she returned to the stars.

We do, in fact, find rock drawings of beings with four fingers at Tiahuanaco. Their age cannot be determined. No one from any of the ages known to us ever saw Tiahuanaco when it was not in ruins.

What secret does this city conceal? What message from other worlds awaits its solution on the Bolivian plateau? There is no plausible explanation for the beginning or the end of this culture. Of course, this does not stop some archaeologists from making the bold and self-confident assertion that the site of the ruins is 3,000 years old. They date this age from a couple of ridiculous little clay figures, which cannot possibly have anything in common with the age of the monolith. Scholars make things very easy for themselves. They stick a couple of old potsherds together, search for one or two adjacent cultures, stick a label on the restored find and—hey, presto!—once again everything fits splendidly into the approved pattern of thought. This method is obviously very much simpler than chancing the idea that an embarrassing technical skill might have existed or the thought of space travelers in the distant past. That would be complicating matters unnecessarily.

Nor must we forget Sacsahuamán! I am not referring here to the fantastic Inca defense works which lie a few feet above present-day Cuzco, nor to the monolithic blocks weighing more than 100 tons, nor to the terrace walls, more than 1,500 feet long and 54 feet wide, in front of which tourists stand and take souvenir snapshots today. I am referring to the unknown Sacsahuamán, which lies a mere half mile or so from the well-known Inca fortress.

Our imagination is unable to conceive what technical resources our forefathers used to extract a monolithic rock of more than 100 tons from a quarry and then transport it and work it in a distant spot. But when we are confronted with a block with an estimated weight of 20,000 tons, our imagination, made rather blasé by the technical achievements of today, is given its severest shock. On the way back from the fortifications of Sacsahuamán, in a crater in the mountainside, a few hundred yards away, the visitor comes across a monstrosity. It is a single stone block the size of a four-story house. It has been impeccably dressed in the most craftsmanlike way; it has steps and ramps and is adorned with spirals and holes. Surely the fashioning of this unprecedented stone block cannot have been merely a bit of leisure activity for the Incas? Surely it is much more likely that it served some as yet inexplicable purpose? To make the solution of the puzzle even more difficult the whole monstrous block stands on its head. So the steps run downward from the roof; the holes point in different directions like the indentations of a grenade; strange depressions, shaped rather like chairs, seem to hang floating in space. Who can imagine that human hands and human endeavor excavated, transported, and dressed this block? What power overturned it?

What titanic forces were at work here?

And to what end?

Still flabbergasted by this stone monstrosity, the visitor finds, barely 900 yards away, rock vitrifications of a kind that ought to be possible only through the melting of stone at extremely high temperatures. The surprised visitor is promptly told that the rock was ground down by glaciers. This explanation is ridiculous. A glacier, like every flowing mass, would logically flow down to one side. This property of matter is hardly likely to have changed just at the time when the vitrifications took place. In any case, it can scarcely be assumed that the glacier flowed down in six different directions over an area of some 18,000 square yards!

Sacsahuamán and Tiahuanaco conceal a great number of prehistorical mysteries for which superficial but quite unconvincing explanations are hawked around. Moreover, sand vitrifications are also found in the Gobi Desert and in the vicinity of old Iraqi archaeological sites. Who can explain why these sand vitrifications resemble those produced by the atomic explosions in the Nevada Desert?

When will something decisive be done to give a convincing answer to the prehistoric puzzles? At Tiahuanaco there are artificial overgrown hills, the "roofs" of which are absolutely level over an area of 4,784 square yards. It seems highly probable that buildings are concealed beneath them. So far no trench has been dug through the chain of hills, no spade is at work to solve the mystery. Admittedly, money is scarce. Yet the traveler often sees soldiers and officers who are obviously at a loss for something useful to do. What is wrong with letting a company of soldiers carry out excavations under expert supervision?

Money is available for so many other things in the world. Research for the future is of burning importance. As long as our past is undiscovered, one entry in the account for the future remains blank. Cannot the past help us to reach technical solutions which will not have to be found for the first time because they already existed in antiquity?

If the urge to discover our past is not sufficient incentive to set modern intensive research work in motion, perhaps the slide rule could be usefully employed. So far, at all events, no scientist has been asked to use the most modern apparatus to investigate radiation at Tiahuanaco, Sacsahuamán, the legendary Sodom, or in the Gobi Desert. Cuneiform texts and tablets from Ur, the oldest books of mankind, tell without exception of "gods" who rode in the heavens in ships, of "gods" who came from the stars, possessed terrible weapons, and returned to the stars. Why do we not seek them out, the old "gods"? Our radio-astronomers send signals into the universe to make contact with unknown intelligences. Why don't we first or simultaneously seek the traces of unknown intelligences on our own earth, which is so much closer? For we are not groping blindly in a dark room—the traces are there for all to see.

Some 2,000 years before our era the Sumerians began to record the glorious past of their people. Today we still do not know where this people came from. But we do know that the Sumerians brought with them a superior advanced culture which they forced upon the still semibarbarian Semites. We also know that they always sought their gods on mountain peaks and that if there were no peaks in the regions they inhabited they erected artificial "mountains" on the plains. Their astronomy was incredibly highly developed. Their observatories achieved estimates of the rotation of the moon which differ from present-day estimates by no more than 0.4 seconds. In addition to the fabulous Epic of Gilgamesh, about which I shall have more to say later, they have left us one thing that is quite sensational. On the hill of Kuyunjik (former Nineveh) a calculation was found with the final result in our notation of 195,955,200,000,000. A number with fifteen digits! Our oft-quoted and extensively studied ancestors of Western culture, the Greeks, never rose above the figure 10,000 during the most brilliant period of their civilization. Anything beyond that was simply described as "infinite."

The old cuneiform inscriptions credit the Sumerians with a literally fantastic span of life. Thus the ten original kings ruled for a total of 456,000 years, and the twenty-three kings who had the arduous task of reconstruction after the Flood still managed to hold the reins of government for a total of 24,510 years, 3 months, and 3½ days.

Periods of years that are quite incomprehensible to our way of thinking, although the names of all the rulers exist in long lists, neatly perpetuated on seals and coins. What would happen if here too we dared to take off our blinkers and look at the old things with fresh eyes, the eyes of today?

Let us suppose that foreign astronauts visited the territory of the Sumerians thousands of years ago. Let us assume that they laid the foundations of the civilization and culture of the Sumerians and then returned to their own planet, after giving this stimulus to development. Let us postulate that curiosity drove them back to the scene of their pioneer work every hundred terrestrial years to check the results of their experiment. By the standards of our present-day expectation of life the same astronauts could easily have survived for 500 terrestrial years. The theory of relativity shows that the astronauts would have aged only about forty years during the outward and return flight in a spaceship that had traveled just under the speed of light! Over the centuries the Sumerians would have built towers, pyramids, and houses with every comfort; they would have sacrificed to their gods and awaited their return. And after hundreds of terrestrial years they actually did return to them. "And then came the Flood, and after the Flood kingship came down from heaven once again," it says in a Sumerian cuneiform inscription.

In what form did the Sumerians imagine and depict their "gods"? Sumerian mythology and some Akkadian tablets and pictures provide information about this. The Sumerian "gods" were not anthropomorphic, and every symbol of a god was also connected with a star. Stars are depicted in Akkadian picture tablets as we might draw them today. The only remarkable thing is that these stars are circled by planets of various sizes. How did the Sumerians, who lacked our techniques for observing the heavens, know that a fixed star has planets? There are sketches in which people wear stars on their heads, while others ride on balls with wings. There is one picture that instantly reminds one of a model of an atom: a circle of balls arranged next to one another that radiate, but are not surrounded by rays. If we look at the legacy of the Sumerians with "space eyes," it teems with questions and enigmas beside which the terrors of the deep and the wonders of the heavens pale into insignificance.

Here are only a few curiosities from the same geographical area:

> Drawings of spirals, a rarity 6,000 years ago, at Geoy Tepe.
>
> A flint industry credited with an age of 40,000 years at Gar Kobeh.
>
> Similar finds at Baradostian estimated to be 30,000 years old.
>
> Figures, tombs, and stone implements at Tepe Asiab dated 13,000 years back.
>
> Petrified excrement, possibly not of human origin, found at the same place.
>
> Tools and stone engravers found at Karim Shahir. Flint weapons and tools excavated at Barda Balka. Skeletons of grown men and a child found in the cave of Shandiar. They were dated (by the C-14 method) to about 45,000 B.C.

The list could be considerably enlarged, and every fact would strengthen the assertion that a mixture of primitive men lived in the geographical territory of Sumer about 40,000 years ago. Suddenly, for reasons inexplicable so far, the Sumerians were there with their astronomy, their culture, and their technology.

The conclusions to be drawn from the previous presence on earth of unknown visitors from the universe are still purely speculative. We can imagine that "gods" appeared who collected the semisavage peoples in the region of Sumer around them and transmitted some of their knowledge to them. The figurines and statues that stare at us today from the glass cases of museums show a racial mixture, with goggle eyes, domed foreheads, narrow lips, and generally long, straight noses. A picture that is very difficult to fit into the schematic system of thought and its concept of primitive peoples.

Visitors from the universe in remote antiquity?

In Lebanon there are glasslike bits of rock, called tektites, in which radioactive aluminum isotopes have been discovered.

In Egypt and Iraq there were finds of cut crystal lenses which today can only be made using cesium oxide, in other words an oxide that has to be produced by electrochemical processes.

In Helwan there is a piece of cloth, a fabric so fine that today it could be woven only in a special factory with great technical know-how and experience.

Electric dry batteries, which work on the galvanic principle, are on display in the Baghdad Museum.

In the same place the visitor can see electric elements with copper electrodes and an unknown electrolyte.

In the mountainous Asian region of Kohistan a cave drawing reproduces the exact position of the stars as they actually were 10,000 years ago. Venus and the earth are joined by lines.

Ornaments of smelted platinum were found on the Peruvian plateau.

Parts of a belt made of aluminum lay in a grave at Yungjen, China.

At Delhi there is an ancient pillar made of iron that is not destroyed by phosphorus, sulphur, or the effects of the weather.

This strange medley of "impossibilities" should make us curious and uneasy. By what means, with what intuition, did the primitive cave dwellers manage to draw the stars in their correct positions? From what precision workshop did the cut crystal lenses come? How could anyone smelt and model platinum, since platinum begins to melt only at 1,800° C? And how did the ancient Chinese make aluminum, a metal which has to be extracted from bauxite under very great chemico-technical difficulties?

Impossible questions, to be sure, but does that mean that we should not ask them? Since we are not prepared to accept or admit that there was a higher culture or an equally perfect technology before our own, all that is left is the hypothesis of a visit from space! As long as archaeology is conducted as it has been thus far, we shall never have a chance to discover whether our dim past was really dim and not perhaps quite enlightened.

A Utopian archaeological year is due, during which archaeologists, physicists, chemists, geologists, metallurgists, and all the corresponding branches of these sciences ought to concentrate their efforts on one single question: Did our forefathers receive visits from outer space?

For example, a metallurgist would be able to tell an archaeologist quickly and concisely how complicated the production of aluminum is. Is it not conceivable that a physicist might instantly recognize a formula in a rock drawing? A chemist with his highly developed apparatus might be able to confirm the assumption that obelisks were extracted from the rock by wetting wooden wedges or using unknown acids. The geologist owes us a whole series of answers to questions about what is of significance in certain Ice Age deposits. The team for a Utopian archaeological year would naturally include a group of divers who would investigate the Dead Sea for radioactive traces of an atomic explosion over Sodom and Gomorrha.

Why are the oldest libraries in the world secret libraries? What are people really afraid of? Are they worried that the truth, protected and concealed for so many thousands of years, will finally come to light?

Research and progress cannot be held back. For 4,000 years the Egyptians considered their "gods" to be real beings. In the Middle Ages we still killed "witches" in our burning ideological zeal. The belief of the ancient Greeks that they could tell the future from a goose's entrails is as out of date today as the conviction of ultraconservatives that nationalism still has the slightest importance.

We have a thousand and one past errors to correct. The self-assurance that is feigned is threadbare and is really only an acute form of stubbornness. At the conference tables of orthodox scientists the delusion still prevails that a thing must be proved before a "serious" person may—or can—concern himself with it.

In the past the man who put forward a brand-new idea had to count on being despised and persecuted by the church and his colleagues. Things must have become easier, one thinks. There are no more anathemas, and fires at the stake are no longer lighted. The snag is that the methods of our time are less spectacular, but they are hardly less obstructive to progress. Now everything is more "civilized" and there is much less fuss. Theories and intolerably audacious ideas are hushed up or dismissed by such killer phrases as:

It's against the rules! (Always a good one!)

It's not classical enough! (Bound to impress.)

It's too revolutionary! (Unequaled in its deterrent effect!)

The universities won't go along with that! (Convincing!)

Others have already tried that! (Of course. But were they successful?)

We can't see any sense in it! (And that's that!)

That hasn't been proved yet! (*Quod erat demonstrandum!*)

Five hundred years ago a scientist cried out in the law courts, "Common sense must tell anyone that the earth cannot possibly be a ball, otherwise the people on the lower half would fall into the void!"

"Nowhere in the Bible," asserted another, "does it say that the earth revolves around the sun. Consequently every such assertion is the work of the devil!"

It seems as if narrow-mindedness was always a special characteristic when new worlds of ideas were beginning.

But on the threshold of the twenty-first century the research worker should be prepared for fantastic realities. He should be eager to revise laws and knowledge which were considered sacrosanct for centuries but are nevertheless called in question by new knowledge. Even if a reactionary army tries to damn up this new intellectual flood, a new world must be conquered in the teeth of all the unteachable, in the name of truth and reality. Anyone who spoke about satellites in scientific circles twenty years ago was committing a kind of academic suicide. Today artificial heavenly bodies, namely satellites, revolve around the sun; they have photographed Mars and landed smoothly on the moon and Venus, radioing first-class photographs of the unknown landscape back to earth with their (tourist) cameras. When the first such photos were radioed to earth from Mars in the spring of 1958, the strength used was 0.000,000,000,000,000,01 watts, an almost incredibly weak amount of current.

Yet *nothing* is incredible any longer. The word "impossible" should have become literally impossible for the modern scientist. Anyone who does not accept this today will be crushed by the reality tomorrow. So let us stick tenaciously to our theory, according to which astronauts from distant planets visited the earth thousands of years ago. We know that our ingenuous and primitive forefathers did not know what to make of the astronauts' superior technology. They worshiped the astronauts as "gods" who came from other stars, and the astronauts had no choice but patiently to accept their adoration as divinities—a homage, incidentally, for which our astronauts on unknown planets must be quite prepared.

Some parts of our earth are still inhabited by primitive peoples to whom a machine gun is a weapon of the devil. In that case a jet aircraft may well be an angelic vehicle to them. And a voice coming from a radio set might seem to be the voice of a god. These last primitive peoples, too, naïvely hand down from generation to generation in their sagas their impressions of technical achievements that we take for granted. They still scratch their divine figures and their wonderful ships coming from heaven on cliffs and cave walls. In this way these savage peoples have actually preserved for us what we are seeking today.

Cave drawings in Kohistan, France, North America, and Southern Rhodesia, in the Sahara and Peru, as well as Chile, all contribute to our theory. Henri Lhote, a French scholar, discovered at Tassili, in the Sahara, several hundred walls painted with many thousands of pictures of animals and men, including figures in short elegant coats. They carry sticks and indefinable chests on the sticks. Next to the animal paintings we are astonished by a being in a kind of diver's suit. The great god Mars—so Lhote christened him—was originally more than 18 feet high; but the "savage" who bequeathed the drawing to us can scarcely have been as primitive as we should like him to be if everything is to fit neatly into the old pattern of thought. After all, the "savage" obviously used a scaffolding to be able to draw in proportion like that, for there have been no shifts in ground level in these caves during the last few millennia. Without overstretching my imagination, I got the impression that the great god Mars is depicted in a space or diving suit. On his heavy, powerful shoulders rests a helmet which is connected to his torso by a kind of joint. There are a number of slits on the helmet where mouth and nose would normally be. One would readily believe that it was the result of chance or even in the pictorial imagination of the prehistoric "artist" if this picture were unique. But there are several of these clumsy figures with the same equipment at Tassili, and very similar figures have also been found on rock faces in the United States, in the Tulare region of California.

I should like to be generous, and I am willing to postulate that the primitive artists were unskilled and portrayed the figures in this rather crude way because it was the best they could do. But in that case how could the same primitive cave dwellers depict animals and normal human beings to perfection? It seems more credible to me to assume that the "artists" were perfectly capable of drawing what they actually saw. In Inyo County, California, a geometrical figure in a cave drawing is recognizable—without overstraining the imagination—as a normal slide rule in a double frame. The archaeological opinion is that the drawing shows figures of the gods.

An animal of unknown species with gigantic upright horns on its head appears on a pottery vessel found at Siyak in Iran. Why not? But both horns display five spirals to left and right. If you imagine two rods with large porcelain insulators, that is roughly what this drawing looks like. What do the archaeologists say to that? Quite simply that they are symbols of a god. Gods are of great value. People explain a great deal—certainly everything that is unexplained—by referring to their unknowableness and supernaturalness. In this world of the undemonstrable they can live in peace. Every figurine that is found, every artifact that is put together, every figure that can be restored from fragments—they are all instantly associated with some ancient religion or other. But if an object cannot be fitted into any of the existing religions, even forcibly, some new crackpot old cult is rapidly conjured up—like a rabbit out of a top hat! The sum works out once again.

But what if the frescoes, at Tassili or in the United States or in France, actually reproduce what the primitive peoples saw? What should we say if the spirals on the rods really depicted antennae, just as the primitive peoples had seen them on the unfamiliar gods? Isn't it possible that things which ought not to exist do in fact exist? A "savage," who nevertheless was skillful enough to execute wall paintings, cannot really have been so savage. The wall drawing of the White

Lady of Brandenberg (South Africa) could be a twentieth-century painting. She wears a short-sleeved pullover, closely fitting breeches, and gloves, garters and slippers. The lady is not alone; behind her stands a thin man with a strange prickly rod in his hand and wearing a very complicated helmet with a kind of visor. This would be accepted as a modern painting without hesitation, but the snag is that we are dealing with a cave drawing.

All the gods who are depicted in cave drawings in Sweden and Norway have uniform indefinable heads. The archaeologists say that they are animal heads. Yet isn't there something rather absurd about worshiping a "god" whom one also slaughters and eats? We often see ships with wings and even more frequently typical antennae.

Figures in bulky suits occur again in Val Camonica (Brescia, Italy), and, annoyingly enough, they also have horns on their heads. I am not going so far as to claim that the Italian cave dwellers shuttled backward and forward between North America or Sweden, the Sahara and Spain (Ciudad Real), to transmit their illustrative talents and ideas. Yet the awkward question is left hanging in the air—why did the primitive people create figures in bulky suits with antennae on their heads independently of each other?

I would not waste a word on these unexplained oddities if they existed in only one place in the world. But they are found almost everywhere.

As soon as we look at the past with our present-day gaze and use the fantasy of our technological age to fill up the gaps in it, the veils that shroud the darkness begin to lift. In the next chapter, a study of ancient holy books will help me to make my theory such a credible reality that in the long run the investigators of our past will no longer be able to evade the revolutionary questions.

4

Was God an Astronaut?

THE Bible is full of secrets and contradictions.

Genesis, for example, begins with the creation of the earth, which is reported with absolute geological accuracy. But how did the chronicler know that minerals preceded plants and plants preceded animals?

"And God said, Let us make man in our image, after our likeness," we read in Genesis 1:26.

Why does God speak in the plural? Why does he say "us," not "me," why "our," and not "my"? One would think that the one and only God ought to address mankind in the singular, not in the plural.

"And it came to pass, when men began to multiply on the face of the earth, and daughters were born unto them, that the sons of God saw the daughters of men that they were fair; and they took them wives of all which they chose" (Genesis 6:1–2).

Who can tell us what sons of God took the daughters of men to wife? Ancient Israel had a single sacrosanct God. Where do the "sons of God" come from?

"There were giants in the earth in those days; and also

after that, when the sons of God came in unto the daughters of men, and they bare children to them, the same became mighty men which were of old, men of renown" (Genesis 6:4).

Once again we have the sons of God, who interbreed with human beings. Here, too, we have the first mention of giants. "Giants" keep on cropping up in all parts of the globe: in the mythology of East and West, in the sagas of Tiahuanaco and the epics of the Eskimos. "Giants" haunt the pages of almost all ancient books. So they must have existed. What sort of creatures were they, these "giants"? Were they our forefathers, who built the gigantic buildings and effortlessly manhandled the monoliths, or were they technically skilled space travelers from another star? One thing is certain. The Bible speaks of "giants" and describes them as "sons of God," and these "sons of God" breed with the daughters of men and multiply.

We are given a very exciting and detailed account of the catastrophe of Sodom and Gomorrah in Genesis 19:1–28.

Two angels came to Sodom in the evening when father Lot was sitting near the town gate. Obviously Lot was expecting these "angels," who soon proved to be men, because he recognized them at once and hospitably invited them to spend the night in his house. The men of the town, says the Bible, wanted to "know" the strangers. But the two strangers were able to dispel the local playboys' sexual lust with a single gesture. They smote the mischief-makers with blindness.

According to Genesis 19:12–14, the "angels" told Lot to take his wife, sons, daughters, sons-in-law, and daughters-in-law out of the town with all speed, for, they warned him, it would be destroyed very soon. The family was unwilling to believe this strange warning and took the whole thing for one of father Lot's bad jokes. And Genesis continues:

"And when the morning arose, then the angels hastened

Lot, saying, Arise, take thy wife, and thy two daughters, which are here; lest thou be consumed in the iniquity of the city. And while he lingered, the men laid hold upon his hand, and upon the hand of his wife, and upon the hand of his two daughters; the Lord being merciful unto him: and they brought him forth, and set him without the city. And it came to pass, when they had brought them forth abroad, that he said, Escape for thy life; look not behind thee, neither stay thou in the plain; escape to the mountain, lest thou be consumed. . . . Haste thee, escape thither; for I cannot do anything till thou be come thither."

According to this report, there is no doubt that the two strangers, the "angels," possessed a power unknown to the inhabitants. The suggestive urgency, the speed with which they drove the Lot family on, also makes us think. When father Lot tarried, they pulled him along by the hands. They had to get away in a matter of minutes. Lot, they ordered, must go into the mountains and he must not turn around. Nevertheless, Lot does not seem to have had unlimited respect for the "angels," because he keeps on making objections: ". . . I cannot escape to the mountain, lest some evil take me, and I die." A little later the angels say that they cannot do anything for him if he does not go with them.

What actually happened at Sodom? We cannot imagine that almighty God is tied down to a timetable. Then why were his "angels" in such a hurry? Or was the destruction of the city by some power or other fixed to the very minute? Had the countdown already begun and did the "angels" know about it? In that case the moment of destruction would obviously have been imminent. Was there no simpler method of bringing the Lot family to safety? Why did they have to go into the mountains at all costs? And why on earth should they be forbidden to look around again?

Admittedly these are awkward questions about a serious matter. But since the dropping of two atomic bombs on

Japan, we know the kind of damage such bombs cause and that living creatures exposed to direct radiation die or become incurably ill. Let us imagine for a moment that Sodom and Gomorrah were destroyed according to plan, *i.e.,* deliberately, by a nuclear explosion. Perhaps—let us speculate a little further—the "angels" simply wanted to destroy some dangerous fissionable material and at the same time to make sure of wiping out a human brood they found unpleasant. The time for the destruction was fixed. Those who were to escape it—such as the Lot family—had to stay a few miles from the center of the explosion in the mountains, for the rock faces would naturally absorb the powerful dangerous rays. And—we all know the story—Lot's wife turned around and looked straight at the atomic sun. Nowadays no one is surprised that she fell dead on the spot. "Then the Lord rained upon Sodom and upon Gomorrah brimstone and fire. . . ."

And this is how the account of the catastrophe ends (Genesis 19:27–28):

"And Abraham got up early in the morning to the place where he stood before the Lord: And he looked toward Sodom and Gomorrah, and toward all the land of the plain, and beheld, and, lo, the smoke of the country went up as the smoke of a furnace."

We may be as religious as our fathers, but we are certainly less credulous. With the best will in the world we cannot imagine an omnipotent, ubiquitous, infinitely good God who is above all concepts of time and yet does not know what is going to happen. God created man and was satisfied with his work. However, he seems to have repented of his deed later, because this same creator decided to destroy mankind. It is also difficult for enlightened children of this age to think of an infinitely good Father who gives preference to "favorite children," such as Lot's family, over countless others. The Old Testament gives some impressive descriptions in which

God alone or his angels fly straight down from heaven making a tremendous noise and issuing clouds of smoke. One of the most original descriptions of such incidents comes to us from the prophet Ezekiel:

"Now it came to pass in the thirtieth year, in the fourth month, in the fifth day of the month, as I was among the captives by the river of Chebar, that the heavens were opened . . . And I looked, and, behold, a whirlwind came out of the north, a great cloud, and a fire infolding itself, and a brightness was about it, and out of the midst thereof as the color of amber, out of the midst of the fire. Also out of the midst thereof came the likeness of four living creatures. And this was their appearance; they had the likeness of a man. And every one had four faces, and every one had four wings. And their feet were straight feet; and the sole of their feet was like the sole of a calf's foot: and they sparkled like the color of burnished brass."

Ezekiel gives precise details of the landing of this vehicle. He describes a craft that comes from the north, emitting rays and gleaming and raising a gigantic cloud of desert sand. Now the God of the Old Testament was supposed to be omnipotent. Then why does this almighty God have to come hurtling up from a particular direction? Cannot he be anywhere he wants without all this noise and fuss?

Let us follow Ezekiel's eyewitness account a little further:

"Now as I beheld the living creatures, behold one wheel upon the earth by the living creatures, with his four faces. The appearance of the wheels and their work was like unto the color of a beryl: and they four had one likeness: and their appearance and their work was as it were a wheel in the middle of a wheel. When they went they went upon their four sides: and they turned not as they went. As for their rings, they were so high that they were dreadful; and their rings were full of eyes round about them four. And when the living creatures went, the wheels went by them: and when the living creatures were lifted up from the earth, the wheels were lifted up."

The description is astonishingly good. Ezekiel says that each wheel was in the middle of another one. An optical illusion! To our present way of thinking what he saw was one of those special vehicles the Americans use in the desert and swampy terrain. Ezekiel observed that the wheels rose from the ground simultaneously with the winged creatures. He was quite right. Naturally the wheels of a multipurpose vehicle, say an amphibious helicopter, do not stay on the ground when it takes off.

More from Ezekiel: "Son of man, stand upon thy feet, and I will speak unto thee."

The narrator heard this voice and hid his face in the ground in fear and trembling. The strange apparitions addressed Ezekiel as "son of man" and wanted to talk to him.

The account goes on: ". . . and I heard behind me a voice of a great rushing, saying, Blessed be the glory of the Lord from his place. I heard also the noise of the wings of the living creatures that touched one another, and the noise of the wheels over against them, and a noise of a great rushing."

In addition to his precise description of the vehicle, Ezekiel also noted the noise the novel monstrosity made when it left the ground. He likens the din made by the wings and the wheels to a "great rushing." Surely this suggests that this is an eyewitness account? The "gods" spoke to Ezekiel and told him that it was his task to restore law and order to the country. They took him with them in their vehicle and confirmed that they had not yet forsaken the country. This experience made a strong impression on Ezekiel, for he never tires of describing the weird vehicle. On three more occasions he says that each wheel was in the middle of a wheel and that the four wheels could go "on their four sides . . . and turned not as they went." And he was particularly impressed by the fact that the whole body of the vehicle, the backs, the hands, the wings, and even the wheels were "full of eyes." The "gods" reveal the purpose and goal of their journey to the chronicler later when they tell him that he is living in the midst of a "rebellious house" which has eyes to see and sees not, ears to hear and hears not. Once he has been enlightened about his countrymen, there follow, as in all descriptions of such landings, advice and directions for law and order, as well as hints for creating a proper civilization. Ezekiel took the task very seriously and handed on the instructions of the "gods."

Once again we are confronted with all kinds of questions. Who spoke to Ezekiel? What sort of beings were they?

They were certainly not "gods" in the traditional sense of the word, or they would not have needed a vehicle to move from one place to another. This kind of locomotion seems to me to be quite incompatible with the idea of an almighty God.

In this connection, there is another technical invention in the Book of Books, which is worthwhile examining impartially.

In Exodus 15:10, Moses relates the exact instructions which "God" gave for building the Ark of the Covenant. The directions are given to the very inch, how and where staves and rings are to be fitted, and from what alloy the metals are to be made. The instructions were meant to ensure that everything was carried out exactly as "God" wanted it. He warned Moses several times not to make any mistakes.

"And look that thou make them after their pattern, which was shewed thee in the mount" (Exodus 25:40).

"God" also told Moses that he would speak to him from the mercy seat. No one, he told Moses, should come close to the Ark of the Covenant, and he gave precise instructions about the clothing to be worn and the footwear appropriate when transporting it. In spite of all this care there was a slipup (2 Samuel 6:2). David had the Ark of the Covenant moved, and Uzzah helped to drive the cart it was in. When passing cattle shook and threatened to overturn the Ark, Uzzah grabbed hold of it. He fell dead on the spot, as if struck by lightning.

Undoubtedly the Ark was electrically charged! If we reconstruct it today according to the instructions handed down by Moses, an electric conductor of several hundred volts is produced. The border and golden crown would have served to charge the condenser which was formed by the gold plates and a positive and negative conductor. If, in addition, one of the two cherubim on the mercy seat acted as a magnet, the loudspeaker—perhaps even a kind of set for communication between Moses and the spaceship—was perfect. The details of the construction of the Ark of the Covenant can be read in the Bible in their entirety. Without actually consulting Exodus, I seem to remember that the Ark was often surrounded by flashing sparks and that Moses made use of this "transmitter" whenever he needed help and advice. Moses heard the voice of his Lord, but he never saw him face to face. When he asked him to show himself to him on one occasion, his "God" answered:

"Thou canst not see my face: for there shall no man see me and live. And the Lord said, Behold, there is a place by me, and thou shalt stand upon a rock: And it shall come to pass, while my glory passeth by, that I will put thee in a clift of the rock, and will cover thee with my hand while I pass by: And I will take away mine hand, and thou shalt see my back parts: but my face shall not be seen" (Exodus 33:20–23).

There are some astonishing similarities in old texts. On the fifth tablet of the Epic of Gilgamesh, which is of Sumerian origin and much older than the Bible, we find virtually the same sentence:

"No mortal comes to the mountain where the gods dwell. He who looks the gods in the face must die."

In other ancient books which hand down stages in the history of mankind, we find very similar statements. Why did the "gods" not want to show themselves face to face? Why did they not let their masks fall? What were they afraid of? Or does the whole account in Exodus come from the Epic of Gilgamesh? Even that is possible. After all, Moses is supposed to have been brought up in the Egyptian royal household. Perhaps he had access to the library or acquired knowledge of ancient secrets during those years.

Perhaps we ought to query our Old Testament dating, too, because there is a good deal to support the fact that David, who lived much later, fought with a giant with six fingers and six toes in his day (2 Samuel 21:18–22). We must also consider the possibility that all the ancient histories, sagas, and narratives were collected and compiled in one spot and later found their way to different countries in the form of copies and somewhat garbled versions.

The finds during recent years near the Dead Sea (the Qumran texts) provide a valuable and astonishing amplification of the biblical Book of Genesis. Once again several hitherto unknown texts mention heavenly chariots, sons of heaven, wheels, and the smoke which the flying apparitions emitted. In the Moses Apocalypse (Chapter 33) Eve looked up to heaven and saw a chariot of light traveling there; it was drawn by four shining eagles. No terrestrial being could have described its magnificence, it says in Moses. Finally the chariot drove up to Adam, and smoke came out from between the wheels. This story, incidentally, does not tell us much that is new. Nevertheless, chariots of light, wheels, and smoke were spoken of as magnificent apparitions as early as and in connection with Adam and Eve.

A fantastic event was deciphered in the Lamech scroll. As the scroll is only fragmentarily preserved, sentences and whole paragraphs of the text are missing. However, what remains is curious enough to be worth retelling.

This tradition says that one fine day Lamech, Noah's father, came home and was surprised to find a boy who, judging by his appearance, was quite out of place in the family. Lamech reproached his wife Bat-Enosh and claimed that the child was not his. Then Bat-Enosh swore by all that was holy that the seed came from him, father Lamech, and not from a soldier or a stranger or one of the "sons of heaven." (In parenthesis we may ask: What sort of "sons of heaven" was Bat-Enosh talking about? At all events, this family drama took place before the Flood.) Nevertheless, Lamech did not believe his wife's protestations and, feeling very upset, went to ask his father Methuselah for advice. On his arrival, he related the family story that was so depressing to him. Methuselah listened to it, reflected, and went off himself to consult the wise Enoch. The cuckoo in the family nest was causing so much trouble that the old man accepted the hardships of the long journey. The question of the little boy's origin had to be cleared up. So Methuselah described how a boy had appeared in his son's family who looked much more like a son of heaven than a man. His eyes, hair, skin, and whole being were unlike those of the rest of the family.

Enoch listened to the story and sent old Methuselah on his way with the extremely worrying news that a great judgment would come upon the earth and mankind and that all "flesh" would be destroyed because it was sordid and dissolute. But the strange boy of whom the family was suspicious had been chosen as the progenitor of those who should survive the great universal judgment. Therefore he should order his son Lamech to call the child Noah. Methuselah journeyed home and told his son Lamech what was in store for them all. What could Lamech do but recognize the unusual child as his own and give him the name of Noah!

The astonishing thing about this family story is the information that Noah's parents were told about the coming Flood and that even grandfather Methuselah was forewarned of the terrible event by the same Enoch who soon afterward, according to tradition, disappeared forever in a fiery heavenly chariot.

Does not this seriously pose the question whether the human race is not an act of deliberate "breeding" by unknown beings from outer space? Otherwise what can be the sense of the constantly recurring fertilization of human beings by giants and sons of heaven, with the consequent extermination of unsuccessful specimens? Seen in this light, the Flood becomes a preconceived project by unknown beings with the intention of exterminating the human race except for a few noble exceptions. But if the Flood, the course of which is historically proved, was quite deliberately planned and prepared—and that several hundred years before Noah received orders to build the ark—then it can no longer be accepted as a divine judgment.

Today the possibility of breeding an intelligent human race is no longer such an absurd theory. Just as the sagas of Tiahuanaco and the inscription on the pediment of the Gate of the Sun talk about a spaceship which landed the Great Mother on earth so that she could bear children, the old religious scripts, too, never tire of saying that "God" created men in his own image. There are texts which note that it needed several experiments before man finally turned out as successfully as "God" wanted. With the theory of a visit to our earth by unknown intelligences from the cosmos, we could postulate that today we are similarly constituted to those fabulous unidentified beings.

In this chain of evidence, the offerings for which the "gods" asked our ancestors raise curious problems. Their demands were by no means limited to incense and animal sacrifices. The lists of gifts required by the gods often include coins made of alloys which are specified in great detail. In fact, the biggest smelting installations in the ancient East were found at Ezion-geber, consisting of a regular ultramodern furnace with a system of air channels, chimney flues, and openings for specific purposes. Smelting experts of our own day are confronted with the as yet unexplained phenomenon of how copper could have been refined in this prehistoric installation. That was undoubtedly the case, for large deposits of copper sulphate were found in the caves and galleries around Ezion-geber. All these finds are estimated to be at least 5,000 years old!

If our own space travelers happen to meet primitive peoples on a planet one day, they too will presumably seem like "sons of heaven" or "gods" to them. Perhaps our intelligences will be as far ahead of the inhabitants of these unknown and as yet unimagined regions as those fabulous apparitions from the universe were ahead of our primitive ancestors. But what a disappointment if time on this as yet unknown landing place had also been progressing and our astronauts were not greeted as "gods" but laughed at as beings living far behind the times!

5

Fiery Chariots from the Heavens

A SENSATIONAL find was made in the hill of Kuyunjik around the turn of the century. It was a heroic epic of great expressive power engraved on twelve clay tablets, and it belonged to the library of the Assyrian King Ashurbanipal. The epic was written in Akkadian; later a second copy was found that goes back to King Hammurabi.

It is an established fact that the original version of the Epic of Gilgamesh stems from the Sumerians, that mysterious people whose origin we do not know but who left behind the astonishing fifteen-digit number and a very advanced astronomy. It is also clear that the main thread of the Epic of Gilgamesh runs parallel to the biblical Book of Genesis.

The first clay tablet of the Kuyunjik finds relates that the victorious hero Gilgamesh built the wall around Uruk. We read that the "god of heaven" lived in a stately home which contained granaries, and that guards stood on the town walls. We learn that Gilgamesh was a mixture of "god" and man —two-thirds "god," one-third man. Pilgrims who came to Uruk gazed up at him in fear and trembling because they had never seen his like for beauty and strength. In other words, the beginning of the narrative contains the idea of interbreeding between "god" and man yet again.

The second tablet tells us that another figure, Enkidu, was created by the goddess of heaven, Aruru. Enkidu is described in great detail. His whole body was covered with hair; he wore skins, ate grass in the fields, and drank at the same watering place as the cattle. He also disported himself in the tumbling waters.

When Gilgamesh, the king of the town of Uruk, heard about this unattractive creature, he suggested that he should be given a lovely woman so that he would become estranged from the cattle. Enkidu, innocent fellow, was taken in by the king's trick and spent six days and six nights with a semidivine beauty. This little bit of royal pandering leads us to think that the idea of cross-breeding between a demigod and a half-animal was not taken quite as a matter of course in this barbaric world.

The third tablet goes on to tell us about a cloud of dust which came from the distance. The heavens roared, the earth quaked, and finally the "sun god" came and seized Enkidu with mighty wings and claws. We read in astonishment that he lay like lead on Enkidu's body and that the weight of his body seemed to him like the weight of a boulder.

Even if we grant the old storytellers a fertile imagination and discount the additions made by translators and copyists, the incredible thing about the account still remains: How on earth could the old chroniclers have known that the weight of the body becomes as heavy as lead at a certain acceleration? Nowadays we know all about the forces of gravity and acceleration. When an astronaut is pressed back into his seat by a force of several G's at takeoff, it has all been calculated in advance.

But how on earth did this idea occur to the old chroniclers?

The fifth tablet narrates how Gilgamesh and Enkidu set out to visit the abode of the "gods" together. The tower in

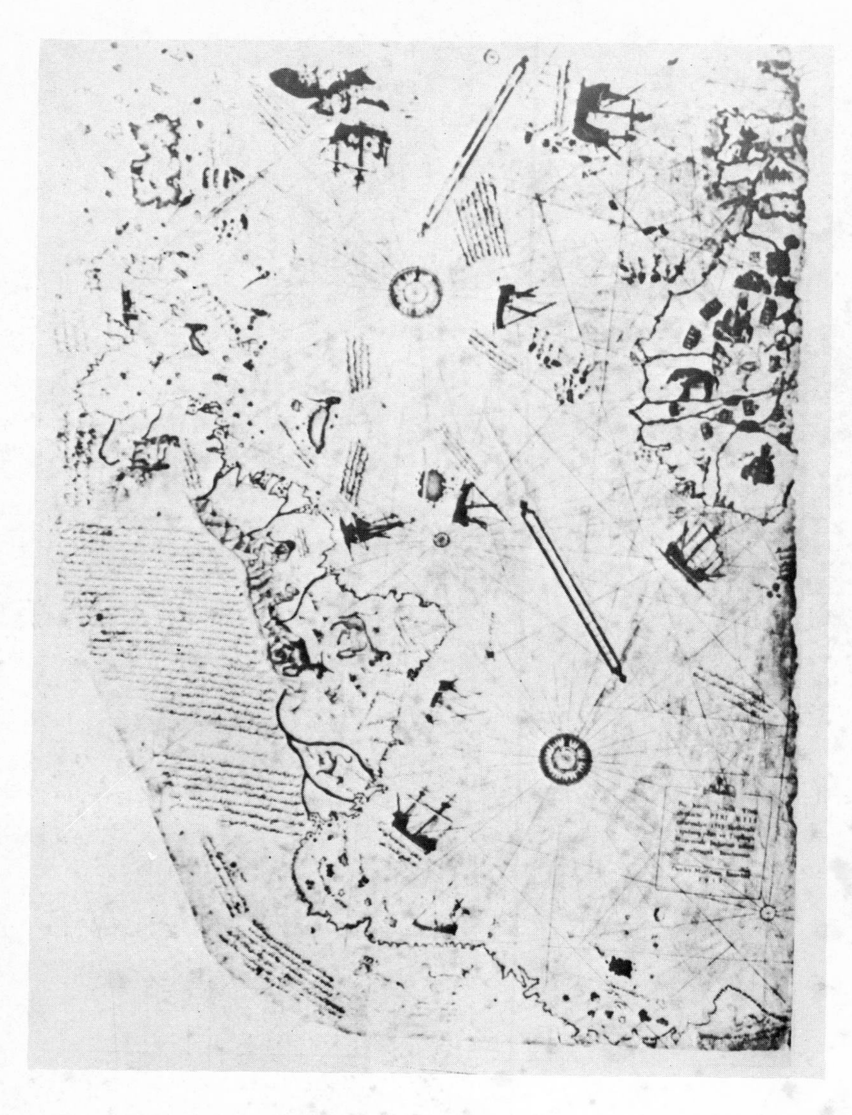

One of the maps found in the Topkapi Palace, Istanbul, in the library of Admiral Piri Reis early in the eighteenth century. It shows the Americas and West Africa. Antarctica, mapped at the bottom, conforms very closely to the land mass under the ice, as revealed by echo-sounding gear. In recorded history it has never been free of ice.

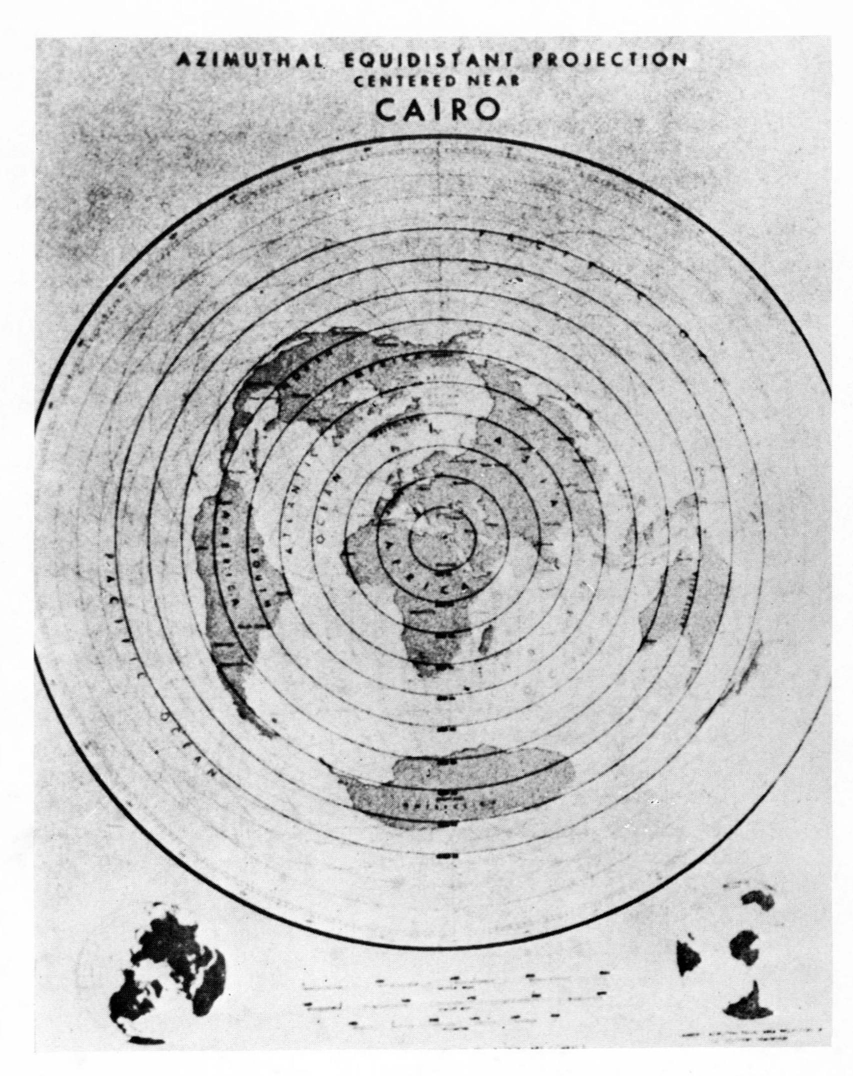

Cartographers projected the Piri Reis map onto a grid using the reference points shown on the map. It then appeared virtually identical with this United States Air Force map of the world on an equidistant projection based on Cairo.

Photo of the earth taken from Apollo 8 shows close similarity to the Piri Reis map. The curious elongated shape of the Americas is particularly striking.

On the Plain of Nazca in Peru appear these strange markings. *Above*: A view of the plain showing that the markings (Inca roads, say the archaeologists) lead nowhere. *Below*: The markings in greater detail.

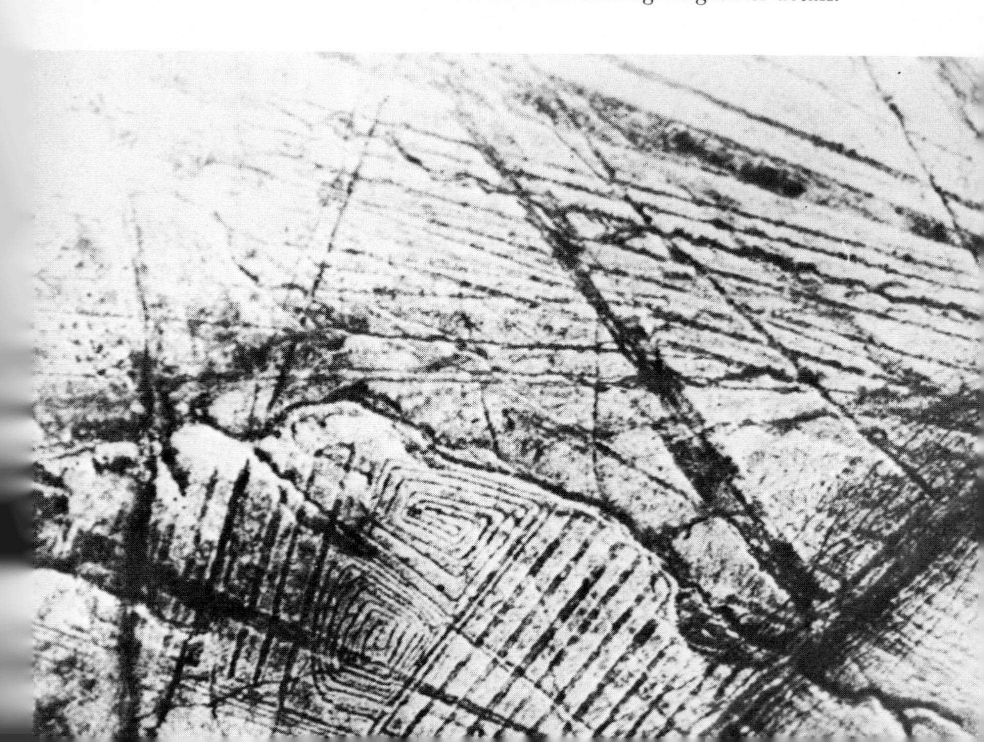

Above: Another of the strange markings on the Plain of Nazca. This is very reminiscent of the aircraft parking areas in a modern airport. *Below*: This huge 820-foot figure above the Bay of Pisco points to the Plain of Nazca. Could this be an aerial direction indicator rather than a symbol of religious significance?

The next five photographs are from South America, a continent teeming with ruins and unexplained puzzles.

Left: Part of a huge monolithic block weighing an estimated 20,000 tons. It can be found at Sacsahuamán, Peru. What was its purpose? What titanic forces turned it upside down?

Below: And what did these great steps lead to? A throne for giants perhaps.

Rock vitrification requires very high temperatures. What caused it in Peru?

Part of the huge terrace walls at Sacsahuamán, Peru. Just look at the incredible accuracy of the jointing. How could primitive people have handled these huge blocks?

El Castillo at Chichén Itzá, Mexico. This has been constructed according to the Mayan calendar. The 91 steps on each side add up to 364, and the final platform gives 365.

In Bolivia near Santa Cruz are long concrete constructions. Could these really be roads for people who did not use the wheel?

Left: This drawing is of the famous funerary crypt in the Temple of Inscriptions in Palenque, Chiapas, Mexico. Could primitive imagination have produced anything so remarkably similar to a modern astronaut in his rocket? The strange markings at the foot of the drawing can only be an indication of the flames and gases coming from the propulsion unit.

Below: American astronauts today take up the same position, their hands on controls, their eyes checking instruments.

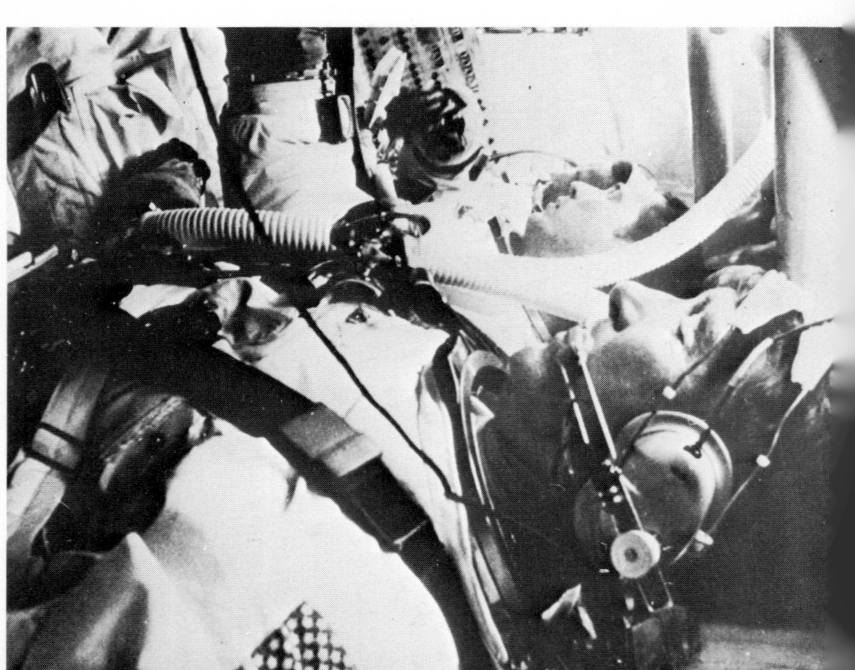

There is no recorded history of Tiahuanaco. On the Gate of the Sun, *above*, carved out of a single 10-ton block, is the representation of a flying god flanked by forty-eight mysterious figures. Legend tells of a golden spaceship which came from the stars.

On the great idol, *left,* is found information about astronomical phenomena covering an immense span of time. And, *below,* once again what primitive people had techniques for handling and accurately jointing such huge blocks of stone, in this case water conduits 6 feet long and 1½ feet wide?

which the goddess Irninis lived could be seen gleaming in the distance long before they reached it. The arrows and missiles which the cautious wanderers rained on the guards rebounded harmlessly. And as they reached the precincts of the "gods," a voice roared at them: "Turn back! No mortal comes to the holy mountain where the gods dwell; he who looks the gods in the face must die."

"Thou canst not see my face, for there shall no man see me and live," it says in Exodus.

On the seventh tablet is the first eyewitness account of a space trip, told by Enkidu. He flew for four hours held in the brazen talons of an eagle. This is how his story goes literally:

"He said to me: 'Look down at the land. What does it look like? Look at the sea. How does it seem to you?' And the land was like a mountain and the sea was like a lake. And again he flew for four hours and said to me: 'Look down at the land. What does it look like? Look at the sea. How does it seem to you?' And the earth was like a garden and the sea like the water channel of a gardener. And he flew higher yet another four hours and spake: 'Look down at the land. What does it look like? Look at the sea. How does it seem to you?' And the land looked like porridge and the sea like a water trough."

In this case some living creature must have seen the earth from a great height. The account is too accurate to have been the product of pure imagination. Who could have possibly said that the land looked like porridge and the sea like a water trough if *some* conception of the globe from above had not existed? Because the earth actually does look like a jigsaw puzzle of porridge and water troughs from a great height.

When the same tablet tells us that a door spoke like a living person, we unhesitatingly identify this strange phenomenon as a loudspeaker. And on the eighth tablet this same

Enkidu, who must have seen the earth from a considerable height, dies of a mysterious disease, so mysterious that Gilgamesh asks whether he may not have been smitten by the poisonous breath of a heavenly beast. But where did Gilgamesh get the idea that the poisonous breath of a heavenly beast could cause a fatal and incurable disease?

The ninth tablet describes how Gilgamesh mourns for the death of his friend Enkidu and decides to undertake a long journey to the gods, because he is obsessed by the idea that he might die of the same disease as Enkidu. The narrative says that Gilgamesh came to two mountains which supported the heavens and that these two mountains arched the gate of the sun. At the gate of the sun he met two giants, and after a lengthy discussion they let him pass because he was two-thirds god himself. Finally Gilgamesh found the garden of the gods, beyond which stretched the endless sea. While Gilgamesh was on his way, the gods warned him twice: "Gilgamesh, whither art thou hurrying? Thou shalt not find the life that thou seekest. When the gods created man, they allotted him to death, but life they retained in their own keeping."

Gilgamesh would not be warned; he wanted to reach Utnapishtim, the father of men, no matter what the dangers. But Utnapishtim lived on the far side of the great sea; no road led to him and no ship flew across it except the sun god's. Braving all kinds of perils Gilgamesh crossed the sea. Then follows his encounter with Utnapishtim, which is described in the eleventh tablet.

Gilgamesh found the figure of the father of men neither bigger nor broader than his own, and he said that they resembled each other like father and son. Then Utnapishtim tells Gilgamesh about his past, strangely enough in the first person.

To our amazement we are given a detailed description of

the Flood. He recounts that the "gods" warned him of the great flood to come and gave him the task of building a boat on which he was to shelter women and children, his relatives, and craftsmen of every kind. The description of the violent storm, the darkness, the rising flood, and the despair of the people he could not take with him has tremendous narrative power even today. We also hear—just as in Noah's account in the Bible—the story of the raven and the dove that were released and how finally, as the waters went down, the boat grounded on a mountain.

The parallel between the stories of the Flood in the Epic of Gilgamesh and the Bible is beyond doubt, and there is not a single scholar who contests it. The fascinating thing about this parallelism is that we are dealing with different omens and different "gods" in this case.

If the account of the Flood in the Bible is a secondhand one, the first-person form of Utnapishtim's narrative shows that a survivor, an eyewitness, was speaking in the Epic of Gilgamesh.

It has been clearly proved that a catastrophic flood did take place in the ancient East some thousands of years ago. Ancient Babylonian cuneiform texts indicate very precisely where the remains of the boat ought to be. And on the south side of Mount Ararat investigators did in fact find three pieces of wood which possibly indicate the place where the ark grounded. Incidentally, the chances of finding the remains of a ship that was mainly built of wood and survived a flood more than 6,000 years ago are extremely remote.

Besides being a first-hand report, the Epic of Gilgamesh also contains descriptions of extraordinary things that could not have been made up by any intelligence living at the time the tablets were written, any more than they could have been devised by the translators and copyists who manhandled the epic over the centuries. For there are facts buried among the descriptions that must have been known to the author of the Epic of Gilgamesh—and we may discover them if we look in the light of present-day knowledge.

Perhaps asking some new questions may throw a little light on the darkness. Is it possible that the Epic of Gilgamesh did not originate in the ancient East at all, but in the Tiahuanaco region? Is it conceivable that descendants of Gilgamesh came from South America and brought the Epic with them? An affirmative answer would at least explain the mention of the Gate of the Sun, the crossing of the sea, and at the same time the sudden appearance of the Sumerians, for as is well known, all the creations of Babylon, which came later, go back to the Sumerians. Undoubtedly the advanced Egyptian culture of the Pharaohs possessed libraries in which the old secrets were preserved, taught, learned, and written down. As has already been mentioned, Moses grew up at the Egyptian court and certainly had access to the venerable library rooms. Moses was a receptive and learned man; indeed he is supposed to have written five of his books himself, although it is still an unsolved puzzle in what language he could have written them.

If we work on the hypothesis that the Epic of Gilgamesh came to Egypt from the Sumerians by way of the Assyrians and Babylonians, and that the young Moses found it there and adapted it for his own ends, then the Sumerian story of the Flood, and not the biblical one, would be the genuine account.

Ought we not to ask such questions? The classical method of research into antiquity has got bogged down and so cannot come to the right unassailable kind of conclusions. It is far too attached to its stereotyped pattern of thought and leaves no scope for the imaginative ideas and speculations which alone could produce a creative impulse.

Many opportunities for research into the ancient East undoubtedly foundered on the inviolability and sacredness of the Bible. People did not dare to ask questions and voice their doubts aloud in the face of this taboo. Even the scholars of the nineteenth and twentieth centuries, ostensibly so enlightened, were still caught in the mental fetters of thousand-year-old errors, because the way back would inevitably have called in question parts of the biblical story. But even very religious Christians must have realized that many of the events described in the Old Testament cannot really be reconciled with the character of a good, great, and omnipresent God. The very man who wants to preserve the religious dogmas of the Bible intact ought to be interested in clarifying who actually educated men in antiquity, who gave them the first rules for a communal life, who handed down the first laws of hygiene, and who annihilated the degenerate stock.

If we think in this way and ask questions like this, it need not mean that we are irreligious. I myself am quite convinced that when the last question about our past has been given a genuine and convincing answer SOMETHING, which I call GOD for want of a better name, will remain for eternity.

Yet the hypothesis that the unimaginable god needed vehicles with wheels and wings to move from place to place, mated with primitive people, and dared not let his mask fall remains an outrageous piece of presumption, as long as it is unsupported by proof. The theologians' answer that God is wise and that we cannot imagine in what way he shows himself and makes his people humble is really dodging our question and is unsatisfactory for that reason. People would like to close their eyes to new realities, too. But the future gnaws away at our past day after day. In the near future, the first men will land on Mars. If there is a single, ancient, long-abandoned edifice there, if there is a single object indicating earlier intelligences, if there is one still recognizable rock drawing to be found, then these finds will shake the foundations of our religions and throw our past into confusion. One single discovery of this kind will cause the greatest revolution and reformation in the history of mankind.

In view of the inevitable confrontation with the future, would it not be more intelligent to use new imaginative ideas when conjuring up our past? Without being unbelieving, we can no longer afford to be credulous. Every religion has an outline, a schema, of its god; it is constrained to think and believe within the framework of this outline. Meanwhile, with the space age, the *intellectual* Day of Judgment comes ever nearer. The theological clouds will evaporate, scattered like shreds of mist. With the decisive step into the universe we shall have to recognize that there are not 2,000,000 gods, not 20,000 sects, not 10 great religions, but only one.

But let us continue to build onto our hypothesis of the Utopian past of humanity. This is the picture so far:

Dim, as yet undefinable ages ago an unknown spaceship discovered our planet. The crew of the spaceship soon found out that the earth had all the prerequisites for intelligent life to develop. Obviously the "man" of those times was no *homo sapiens* but something rather different. The spacemen artificially fertilized some female members of this species, put them into a deep sleep, so ancient legends say, and departed. Thousands of years later the space travelers returned and found scattered specimens of the genus *homo sapiens.* They repeated their breeding experiment several times until finally they produced a creature intelligent enough to have the rules of society imparted to it. The people of that age were still barbaric. Because there was a danger that they might retrogress and mate with animals again, the space travelers destroyed the unsuccessful specimens or took them with them to settle them on other continents. The first communities and the first skills came into being; rock faces and cave walls were painted, pottery was discovered, and the first attempts at architecture were made.

These first men had tremendous respect for the space travelers. Because they came from somewhere absolutely unknown and then returned there again, they were the "gods" to them. For some mysterious reason the "gods" were interested in passing on their intelligence. They took care of the creatures they bred; they wanted to protect them from corruption and preserve them from evil. They wanted to ensure that their community developed constructively. They wiped out the freaks and saw to it that the remainder received the basic requirements for a society capable of development.

Admittedly this speculation is still full of holes. I shall be told that proofs are lacking. The future will show how many of those holes can be filled in. This book puts forward a hypothesis made up of many speculations, therefore the hypothesis must not be "true." Yet when I compare it with the theories enabling many religions to live unassailed in the shelter of their taboos, I should like to attribute a minimal percentage of probability to my hypothesis.

Perhaps it will do some good to say a few words about the "truth." Anyone who believes in a religion and has never been under attack is convinced that he has the "truth." That applies not only to Christians but also to the members of other religious communities, both large and small. Theosophists, theologists, and philosophers have reflected about their teaching, about their master and his teaching; they are convinced that they have found the "truth." Naturally every religion has its history, its promises made by God, its covenants with God, its prophets and wise teachers who have said. . . . Proofs of the "truth" always start from the center of one's own religion and work outward. The result is a biased way of thinking which we are brought up to accept from childhood. Nevertheless generations lived and still do live in the conviction that they possess the "truth."

Somewhat more modestly, I claim that we cannot possess

the "truth." At best we can believe in it. Anyone who really seeks *the* truth cannot and ought not to seek it under the aegis and within the confines of his own religion. If he does so, is not insincerity godfather to a matter which demands the greatest integrity? What is the purpose and goal of life after all? To believe in the "truth" or to seek it?

Even if Old Testament facts can be proved archaeologically in Mesopotamia, those verified facts are still no proof of the religion concerned. If ancient cities, villages, wells, and inscriptions are dug up in a particular area, the finds show that the history of the people who lived there is an actual fact. But they do not prove that the god of that people was the one and only god (and not a space traveler).

Today excavations all over the world show that traditions tally with the facts. But would it occur to a single Christian to recognize the god of the pre-Inca culture as the *genuine* god as the results of excavations in Peru? Quite simply what I mean is that everything, both myth and actual experience, makes up the history of a people. No more. But even that, I claim, is quite a lot.

So anyone who really seeks truth cannot ignore new and bold and as yet unproved points of view simply because they do not fit into his scheme of thought (or belief). Since the question of space travel did not arise a hundred years ago, our fathers and grandfathers could not reasonably have had thoughts about whether our ancestors had visits from the universe. Let us just venture the frightful, but unfortunately possible, idea that our present-day civilization was entirely destroyed in an H-bomb war. Five thousand years later archaeologists would find fragments of the Statue of Liberty in New York. According to our current way of thinking they would be bound to assert that they were dealing with an unknown divinity, probably a fire god (because of the torch) or a sun god (because of the rays around the statue's head).

They would never dare to say that it was a perfectly simple artifact, namely a statue of liberty.

It is no longer possible to block the roads to the past with dogmas.

If we want to set out on the arduous search for the truth, we must all summon up the courage to leave the lines along which we have thought until now and as the first step begin to doubt everything that we previously accepted as correct and true. Can we still afford to close our eyes and stop up our ears because new ideas are supposed to be heretical and absurd?

After all, the idea of a landing on the moon was absurd fifty years ago.

6

Ancient Imagination and Legends, or Ancient Facts?

AS I have previously observed, there were things in antiquity that should not have existed according to current ideas. But my collector's zeal is by no means exhausted with the finds already accumulated.

Why? Because the mythology of the Eskimos also says that the first tribes were brought to the North by "gods" with brazen wings! The oldest American Indian sagas mention a thunderbird who introduced fire and fruit to them. The Mayan legend, the Popol Vuh, tells us that the "gods" were able to recognize everything: the universe, the four cardinal points of the compass, and even the round shape of the earth.

What are the Eskimos doing talking about metal birds? Why do the Indians mention a thunderbird? How are the ancestors of the Mayas supposed to have known that the earth is round?

The Mayas were intelligent; they had a highly developed culture. They left behind not only a fabulous calendar but also incredible calculations. They knew the Venusian year of 584 days and estimated the duration of the terrestrial year at 365.2420 days. (The exact calculation today: 365.2422!) The Mayas left behind them calculations to last for 64,000,000 years. Later inscriptions dealt in units which probably approach 400,000,000 years. The famous Venusian formula could quite plausibly have been calculated by an electronic brain. At any rate, it is difficult to believe that it originated from a jungle people. The Venusian formula of the Mayas runs as follows:

The Tzolkin has 260 days, the terrestrial year 365 days, and the Venusian year 584 days. These figures conceal the possibility of an astonishing division sum. 365 is divisible by 73 five times, and 584 eight times. So the incredible formula takes this form:

(Moon)	$20 \times 13 = 260 \times 2 \times 73 = 37{,}960$
(Sun)	$8 \times 13 = 104 \times 5 \times 73 = 37{,}960$
(Venus)	$5 \times 13 = 65 \times 8 \times 73 = 37{,}960$

In other words, all the cycles coincide after 37,960 days. Mayan mythology claimed that then the "gods" would come to the great resting place.

The religious legends of the pre-Inca peoples say that the stars were inhabited and that the "gods" came down to them from the constellation of the Pleiades. Sumerian, Assyrian, Babylonian, and Egyptian cuneiform inscriptions constantly present the same picture: "gods" came from the stars and went back to them; they traveled through the heavens in fireships or boats, possessed terrifying weapons, and promised immortality to individual men.

It was, of course, perfectly natural for the ancient peoples to seek their gods in the sky and also to give their imagination full rein when describing the magnificence of these incomprehensible apparitions. Yet even if all that is accepted, there are still too many anomalies left.

For example, how did the chronicler of the Mahabharata

know that a weapon capable of punishing a country with a twelve years' drought could exist? And powerful enough to kill the unborn in their mothers' wombs? This ancient Indian epic, the Mahabharata, is more comprehensive than the Bible, and even at a conservative estimate its original core is at least 5,000 years old. It is well worth reading this epic in the light of present-day knowledge.

We shall not be very surprised when we learn in the Ramayana that Vimanas, *i.e.,* flying machines, navigated at great heights with the aid of quicksilver and a great propulsive wind. The Vimanas could cover vast distances and could travel forward, upward and downward. Enviably maneuverable space vehicles!

This quotation comes from the translation by N. Dutt, 1891: "At Rama's behest the magnificent chariot rose up to a mountain of cloud with a tremendous din. . . ." We cannot help noticing that not only is a flying object mentioned again but also that the chronicler talks of a tremendous din.

Here is another passage from the Mahabharata: "Bhima flew with his Vimana on an enormous ray which was as brilliant as the sun and made a noise like the thunder of a storm" (C. Roy, 1889).

Even imagination needs something to start it off. How can the chronicler give descriptions that presuppose at least some idea of rockets and the knowledge that such a vehicle can ride on a ray and cause a terrifying thunder?

In the Samsaptakabadha a distinction is made between chariots that fly and those that cannot fly. The first book of the Mahabharata reveals the intimate history of the unmarried Kunti, who not only received a visit from the sun god but also had a son by him, a son who is supposed to have been as radiant as the sun itself. As Kunti was afraid—even in those days—of falling into disgrace, she laid the child in a little basket and put it in a river. Adhirata, a worthy man of the

Suta caste, fished basket and child out of the water, and brought up the infant.

Really a story that is hardly worth mentioning if it were not so remarkably like the story of Moses! And, of course, there is yet another reference to the fertilization of humans by gods. Like Gilgamesh, Aryuna, the hero of the Mahabharata, undertakes a long journey in order to seek the gods and ask them for weapons. And when Aryuna has found the gods after many perils, Indra, the lord of heaven, with his wife, Sachi, beside him, grants him a very exclusive audience. The two do not meet the valiant Aryuna just anywhere. They meet him in a heavenly war chariot and even invite him to travel in the sky with them.

Certain numerical data in the Mahabharata are so precise that one gets the impression that the author was writing from first-hand knowledge. Full of repulsion, he describes a weapon that could kill all warriors who wore metal on their bodies. If the warriors learned about the effect of this weapon in time, they tore off all the metal equipment they were wearing, jumped into a river, and washed themselves and everything that they had come into contact with very thoroughly. Not without reason, as the author explains, for the weapon made the hair and nails fall out. Everything living, he bemoaned, became pale and weak.

In the eighth book we meet Indra in his heavenly jet chariot again. Out of the whole of mankind he has chosen Yudhisthira as the only one who may enter heaven in his mortal frame. Here, too, the parallel with the stories of Enoch and Elijah cannot be overlooked.

In the same book, in what is perhaps the first account of the dropping of an H-bomb, it says that Gurkha loosed a single projectile on the triple city from a mighty Vimana. The narrative uses words which linger in our memories from eyewitness accounts of the detonation of the first hydrogen bomb at Bikini: white-hot smoke, a thousand times brighter than the sun, rose up in infinite brilliance and reduced the city to ashes. When Gurkha landed again, his vehicle was like a flashing block of antimony. And for the benefit of the philosophers I should mention that the Mahabharata says that time is the seed of the universe.

The Tibetan books Tantyua and Kantyua also mention prehistoric flying machines, which they call "pearls in the sky." Both books expressly emphasize that this knowledge is secret and not for the masses. In the Samarangana Sutradhara whole chapters are devoted to describing airships whose tails spout fire and quicksilver.

The word "fire" in ancient texts cannot mean burning fire, for altogether some forty different kinds of "fire," mainly connected with esoteric and magnetic phenomena, are enumerated. It is hard to believe that the ancient peoples should have known that it is possible to gain energy from heavy metals and how to do so. However, we should not oversimplify and dismiss the old Sanscrit texts as mere myths. The large number of passages from old texts already quoted turns the suspicion that men encountered flying "gods" in antiquity almost into a certainty. We are not going to get any further with the old approach which scholars unfortunately still cling to: "That doesn't exist . . . those are mistakes in translation . . . those are fanciful exaggerations by the author or copyists." We must use a new working hypothesis, one developed from the technological knowledge of our age, to throw light onto the thicket behind which our past lies concealed. Just as the phenomenon of the spaceship in the remote past is explicable, there is also a plausible explanation of the terrible weapons which the gods made use of at least once in those days and which are so frequently described. A passage from the Mahabharata is bound to make us think:

> It was as if the elements had been unleashed. The sun spun round. Scorched by the incandescent heat of the weapon, the world reeled in fever. Elephants were set on fire by the heat

> and ran to and fro in a frenzy to seek protection from the terrible violence. The water boiled, the animals died, the enemy was mown down and the raging of the blaze made the trees collapse in rows as in a forest fire. The elephants made a fearful trumpeting and sank dead to the ground over a vast area. Horses and war chariots were burnt up and the scene looked like the aftermath of a conflagration. Thousands of chariots were destroyed, then deep silence descended on the sea. The winds began to blow and the earth grew bright. It was a terrible sight to see. The corpses of the fallen were mutilated by the terrible heat so that they no longer looked like human beings. Never before have we seen such a ghastly weapon and never before have we heard of such a weapon (C. Roy, 1889).

The story goes on to say that those who escaped washed themselves, their equipment, and their arms, because everything was polluted by the death-dealing breath of the "gods." What does it say in the Epic of Gilgamesh? "Has the poisonous breath of the heavenly beast smitten you?"

Alberto Tulli, formerly keeper of the Egyptian Department in the Vatican Museum, found a fragment of a text from the time of Thutmose III, who lived about 1500 B.C. It relates the tradition that the scribes saw a ball of fire come down from heaven and that its breath had an evil smell. Thutmose and his soldiers watched this spectacle until the ball of fire rose in a southerly direction and disappeared from view.

All the texts quoted date from millennia before our era. The authors lived on different continents and belonged to different cultures and religions. There were no special messengers to spread the news in those days, and intercontinental journeys were not an everyday occurrence. In spite of this, traditions telling almost the same story come from the four corners of the world and from innumerable sources. Did all their authors have the same bee in their bonnet? Were they all haunted by the same phenomenon? It is impossible and incredible that the chroniclers of the Mahabharata, the Bible, the Epic of Gilgamesh, the texts of the Eskimos, the American Indians, the Scandinavians, the Tibetans, and many, many other sources should all tell the same stories—of flying "gods," strange heavenly vehicles, and the frightful catastrophes connected with these apparitions—by chance and without any foundation. They cannot all have had the same ideas all over the world. The almost uniform texts can stem only from facts, *i.e.,* from prehistoric events. They related what was actually there to see. Even if the reporter in the remote past may have exaggerated his story with fanciful trimmings, much as newsmen do today, the fact, the actual incident, still remains at the core of all exclusive accounts, as it does today. And that incident obviously cannot have been invented in so many places in different ages.

Let us make up an example:

A helicopter lands in the African bush for the first time. None of the natives has ever seen such a machine. The helicopter lands in a clearing with a sinister clatter; pilots in battle dress, with crash helmets and machine guns, jump out of it. The savage in his loincloth stands stupefied and uncomprehending in the presence of this thing that has come down from heaven and the unknown "gods" who came with it. After a time the helicopter takes off again and disappears into the sky.

Once he is alone again, the savage has to work out and interpret this apparition. He will tell others who were not present what he saw: a bird, a heavenly vehicle, that made a terrible noise and stank, and white-skinned creatures carrying weapons that spat fire. The miraculous visit is fixed and handed down for all time. When the father tells it to his son, the heavenly bird obviously does not get any smaller, and the creatures that got out of it become weirder, stronger, and more imposing. These and many other embellishments will be added to the story. But the premise for the glorious legend was the actual landing of the helicopter. It did land in the clearing in the jungle and the pilots did climb out of it. From that moment the event is perpetuated in the mythology of the tribe.

Certain things cannot be made up. I should not be ransacking our prehistory for space travelers and heavenly aircraft if accounts of such apparitions appeared in only two or three ancient books. But when in fact nearly all the texts of the primitive peoples all over the globe tell the same story, I feel I must try to explain the objective thrust concealed in their pages.

"Son of man, thou dwellest in the midst of a rebellious house, which have eyes to see, and see not; they have ears to hear, and hear not . . ." (Ezekiel 12:2).

We know that all the Sumerian gods had their counterparts in certain stars. There is supposed to have been a statue to Marduk (Mars), the highest of the gods, that weighed 800 talents of pure gold. If we are to believe Herodotus, that is equivalent to more than 48,000 pounds of gold. Ninurta (Sirius) was judge of the universe and passed sentence on mortal men. There are cuneiform tablets which were addressed to Mars, to Sirius, and to the Pleiades. Time and again Sumerian hymns and prayers mention divine weapons, the form and effect of which must have been completely senseless to the people of those days. A panegyric to Mars says that he made fire rain down and destroyed his enemies with a brilliant lightning flash. Inanna is described as she traverses the heavens, radiating a frightful blinding gleam and annihilating the houses of the enemy. Drawings and even the model of a home have been found resembling a prefabricated atomic bunker: round and massive, with a single strangely framed aperture. From the same period, about 3000 B.C., archaeologists have found a model of a team with chariot and driver, as well as two sportsmen wrestling, all of immaculate craftsmanship. The Sumerians, it has been proved, were masters of applied art. Then why did they model a clumsy bunker, when other excavations at Babylon or Uruk have brought much subtler works to light? Quite recently a whole Sumerian library of about 60,000 clay tablets was found in the town of Nippur, 95 miles south of Baghdad. We now possess the oldest account of the Flood, engraved on a tablet in six colums. Five antediluvian cities are named on the tablets: Eridu, Badtibira, Larak, Sitpar, and Shuruppak. Two of these cities have not yet been discovered. On these tablets, the oldest deciphered to date, the Noah of the Sumerians is called Ziusudra. He is supposed to have lived in Shuruppak and also to have built his ark there. So we now possess an even older description of the Flood than the one in the Epic of Gilgamesh. No one knows whether new finds will not produce still earlier accounts.

The men of the ancient cultures seem to have been almost obsessed with the idea of immortality or rebirth. Servants and slaves obviously lay down voluntarily in the tomb with their masters. In the burial chamber of Shub-At, no less than seventy skeletons lay next to each other in perfect order. Without the least sign of violence, sitting or lying in their brilliantly colored robes, they awaited the death which must have come swiftly and painlessly—perhaps by poison. With unshakable conviction, they looked forward to a new life beyond the grave with their masters. But who put the idea of rebirth into the heads of these heathen peoples?

The Egyptian pantheon is just as confusing. The ancient texts of the people on the Nile also tell of mighty beings who traversed the firmament in boats. A cuneiform text to the sun god Ra runs: "Thou couplest under the stars and the moon, thou drawest the ship of Aten in heaven and on earth like the tirelessly revolving stars and the stars at the North Pole that do not set."

Here is an inscription from a pyramid: "Thou art he who directs the sun ship of millions of years."

Even if the old Egyptian mathematicians were very advanced, it is odd that they should speak of millions of years in connection with the stars and a heavenly ship. What does the Mahabharata say? "Time is the seed of the universe."

In Memphis the god Ptah handed the king two models with which to celebrate the anniversaries of his reign and commanded him to celebrate the said anniversaries for six times a hundred thousand years. When the god Ptah came to give the king the models he appeared in a gleaming heavenly chariot and afterward disappeared over the horizon in it. Today representations of the winged sun and a soaring falcon carrying the sign of eternity and eternal life can still be found on doors and temples at Idfu. There is no known place in the world where such innumerable illustrations of winged symbols of the gods are preserved as in Egypt.

Every tourist knows the Island of Elephantine with the famous Nilometer at Aswan. The island is called Elephantine even in the oldest texts, because it was supposed to resemble an elephant. The texts were quite right—the island does look like an elephant. But how did the ancient Egyptians know that? This shape can be recognized only from an airplane at a great height, for there is no hill offering a view of the island that would prompt anyone to make the comparison.

A recently discovered inscription on a building at Idfu says that the edifice is of supernatural origin. The ground plan was drawn by the deified being Im-Hotep. Now this Im-Hotep was a very mysterious and clever personality—the Einstein of his time. He was priest, scribe, doctor, architect, and philosopher rolled into one. In this ancient world, the age of Im-Hotep, according to archaeologists, the only tools the people could have used for working stone were wooden wedges and copper, neither of which is suitable for cutting up granite blocks. Yet the brilliant Im-Hotep built the step pyramid of Sakkara for his king, who was called Zoser. This 197-foot-high edifice is built with a mastery that Egyptian architects were never quite able to equal afterward. The structure, surrounded by a wall 33 feet high and 1,750 feet long, was called the House of Eternity by Im-Hotep. He had himself buried in it, so that the gods could wake him on their return.

We know that all the pyramids were laid out according to the positions of certain stars. Is not this knowledge a bit embarrassing in view of the fact that we have very little evidence of Egyptian astronomy? Sirius was one of the few stars they took an interest in. But this very interest in Sirius seems rather peculiar, because seen from Memphis, Sirius can be observed only in the early dawn just above the horizon when the Nile floods begin. To fill the measure of confusion to overflowing, there was an accurate calendar in Egypt 4,221 years before our era! This calendar was based on the rise of Sirius (1st Tout = July 19) and gave annual cycles of more than 32,000 years.

Admittedly the old astronomers had plenty of time to observe the sun, the moon, and the stars, year in, year out, until they finally decided that all the stars stand in the same place again after approximately 365 days. But surely it was quite absurd to base the first calendar on Sirius when it would have been easier to use the sun and the moon, besides leading to more accurate results. Presumably the Sirius calendar is a built-up system, a theory of probabilities, because it could never predict the appearance of the star. If Sirius appeared on the horizon at dawn at the same time as the Nile flood, it was pure coincidence. A Nile flood did not happen every year, nor did every Nile flood take place on the same day. In which case, why a Sirius calendar? Is there an old tradition here, too? Was there a text or a promise which was carefully guarded by the priesthood?

The tomb in which a gold necklace and the skeleton of an entirely unknown animal were found probably belonged to King Udimu. Where did the animal come from? How can we explain the fact that the Egyptians had a decimal system already at the beginning of the first dynasty? How did such a highly developed civilization arise at such an early date? Where do objects of copper and bronze originate as early as the beginning of the Egyptian culture? Who gave them their incredible knowledge of mathematics and a ready-made writing?

Before we deal with some monumental buildings which raise innumerable questions, let us take another brief glance at the old texts.

Where did the narrators of *The Thousand and One Nights* get their staggering wealth of ideas? How did anyone come to describe a lamp from which a magician spoke when the owner wished?

What daring imagination invented the "Open, Sesame!" incident in the tale of Ali Baba and the forty thieves?

Of course, such ideas no longer astonish us today, for the television set shows us talking pictures at the turn of a switch. And as the doors of most large department stores open by photocells, even the "Open, Sesame!" incident no longer conceals any special mystery. Nevertheless the imaginative power of the old storytellers was so incredible that the books of contemporary writers of science fiction seem banal in comparison. So it must be that the ancient storytellers had a store of things already seen, known, and experienced ready at hand to spark off their imagination!

In the legendary and sagalike world of intangible cultures which as yet offer to us no fixed points of reference, we are on still shakier ground, and things become even more confusing.

Naturally the Icelandic and Old Norwegian traditions also mention "gods" who travel in the sky. The goddess Frigg has a maidservant called Gna. The goddess sends her handmaid to different worlds on a steed which rises in the air above land and sea. The steed is called "Hoof-thrower," and once, says the saga, Gna met some strange creatures high in the air. In the *Alwislied* different names are given to the earth, the sun, the moon, and the universe depending on whether they are seen from the point of view of men, "gods," giants, or dwarfs. How on earth could people in the dim past arrive at different perceptions of one and the same thing, when the horizon was very limited?

Although the scholar Snorri Sturluson did not write down the Nordic and Old Germanic legends, sagas, and songs until about A.D. 1200, they are known to be some thousands of years old. In these writings the symbol of the world is often described as a disc or a ball—remarkably enough—and Thor, the leader of the gods, is always shown with a hammer, the destroyer. Herbert Kühn supports the view that the word "hammer" means "stone," dates from the Stone Age, and was transferred to bronze and iron hammers only later. Consequently Thor and his hammer symbol must have been very ancient and probably do go back to the Stone Age. Moreover, the word "Thor" in the Indian (Sanscrit) legends is "Tanayitnu"; this could be more or less rendered as "the Thunderer." The Nordic Thor, god of gods, is the lord of the Germanic *Wannen,* who makes the skies unsafe.

When arguing about the entirely new aspects that I introduce into investigation of the past, the objection might be made that it is not possible to compile everything in the ancient traditions that points to heavenly apparitions into a sequence of proofs of prehistoric space travel. But that is not what I am doing. I am simply referring to passages in very ancient texts that have no place in the working hypothesis in use up to the present. I am drilling away at those admittedly awkward spots in which scribes, translators, and copyists could have had no idea of the sciences and their products. I also would be quite prepared to consider the translations wrong and the copies not accurate enough if these same false, fancifully embellished traditions were not accepted in their entirety as soon as they can be fitted into the framework of some religion or other. It is unworthy of a scientific investigator to deny something when it upsets his working hypothesis and accept it when it supports his theory. Imagine the shape my theory would take and the strength it would gain if new translations made with a "space outlook" existed!

To help us patiently forge the chain of our thesis a little further, scrolls with fragments of apocalyptic and liturgical texts were recently found near the Dead Sea. Once again, in the Apocryphical Books of Abraham and Moses, we hear about a heavenly chariot with wheels, which spits fire, whereas similar references are lacking in the Ethiopian and Slavic Book of Enoch.

"Behind the being I saw a chariot which had wheels of fire, and every wheel was full of eyes all around, and on the wheels was a throne and this was covered with fire that flowed around it" (Apocryphal Book of Abraham 18:11–12).

According to Gershom Scholem's explanation, the throne and chariot symbolism of the Jewish mystics corresponded roughly to that of the Hellenistic and early Christian mystics when they talk about *pleroma* (abundance of light). That is a respectable explanation, but can it be accepted as scientifically proved? May we simply ask what would be the case if some people had really seen the fiery chariot that is described over and over again? A secret script was used very frequently in the Qumran scrolls; among the documents in the fourth cave different kinds of characters alternate in one and the same astrological work. An astronomical observation bears the title: "Words of the judicious one which he has addressed to all sons of the dawn."

But what is the crushing and convincing objection to the possibility that real fiery chariots were described in the ancient texts? Surely not the vague and stupid assertion that fiery chariots cannot have existed in antiquity! Such an answer would be unworthy of the men I am trying to force to face new alternatives with my questions. Lastly, it is by no means so long ago that reputable scholars said that no stones (meteors) could fall from the sky, because there were no stones in the sky. Even nineteenth-century mathematicians came to the conclusion—convincing in their day—that a railway train would not be able to travel faster than 21 miles an hour because if it did the air would be forced out of it and the passengers would suffocate. Less than a hundred years ago it was "proved" that an object heavier than air would never be able to fly.

A review in a reputable newspaper classed Walter Sullivan's book *We Are Not Alone* as science fiction and said that even in the most distant future it would be quite impossible to reach, say, Epsilon Eridani or Tau Ceti; even the effect of a shift in time or deep-freezing the astronauts could never overcome the barriers of the inconceivable distances.

It is a good thing that there were always enough bold visionaries oblivious to contemporary criticism in the past. Without them there would be no worldwide railway network today, with trains traveling at 124 miles an hour and more. (N.B.: Passengers die at more than 21 miles per hour!) Without them there would be no jet aircraft today, because they would certainly fall to the ground. (Things that are heavier than air cannot fly!) And there would be no moon rockets. (Man cannot leave his own planet!) There are so many, many things that would not exist but for the visionaries!

A number of scholars would like to stick to the so-called realities. In so doing they are too ready and willing to forget that what is reality today may have been the Utopian dream of a visionary yesterday. We owe a considerable number of all the epoch-making discoveries that our age thinks of as realities to lucky chances, not to steady systematic research. And some of them stand to the credit of the "serious visionaries" who overcame restricting prejudice with their bold speculations. For example, Heinrich Schliemann accepted Homer's *Odyssey* as more than stories and fables and discovered Troy as a result.

We still know too little about our past to be able to make a definite judgment about it. New finds may solve unprecedented mysteries; the reading of ancient narratives is capable of turning whole worlds of realities upside down. Incidentally, it is obvious to me that more old books were destroyed than are preserved. There is supposed to have been a book in South America that contained all the wisdom of antiquity; it is reputed to have been destroyed by the sixty-third Inca ruler, Pachacuti IV. In the library of Alexandria 500,000 volumes belonging to the learned Ptolemy Soter contained all the traditions of mankind; the library was partly destroyed by the Romans, and the rest was burned on the orders of Caliph Omar centuries later. An incredible thought that invaluable and irreplaceable manuscripts were used to heat the public baths of Alexandria!

What became of the library of the temple at Jerusalem? What became of the library of Pergamon, which is supposed to have housed 200,000 works? When the Chinese Emperor Chi-Huang ordered the destruction of a mass of historical, astronomical, and philosophical books for political reasons in 214 B.C., what treasures and secrets went with them? How many texts did the converted Paul cause to be destroyed at Ephesus? And we cannot even imagine the enormous wealth of literature about all branches of knowledge that has been lost to us owing to religious fanaticism. How many thousands of irretrievable writings did monks and missionaries burn in South America in their blind religious zeal?

That happened hundreds and thousands of years ago. Has mankind learned anything as a result? Only half a century ago Hitler had books burned in the public squares, and as recently as 1966 the same thing happened in China during Mao's kindergarten revolution. Thank heavens that today books do not exist in single copies, as in the past.

The texts and fragments still available transmit a great deal of knowledge from the remote past. In all ages the sages of a nation knew that the future would always bring wars and revolutions, blood and fire. Did this knowledge perhaps lead these sages to hide secrets and traditions from the mob in the colossal buildings of their period or to preserve them from possible destruction in a safe place? Have they "hidden" information or accounts in pyramids, temples, and statues, or bequeathed them in the form of ciphers so that they would withstand the ravages of time? We certainly ought to test the idea, for farsighted contemporaries of our own day have acted in this way—for the future.

In 1965 Americans buried in the soil of New York two time capsules so constituted that they could withstand the very worst that this earth could offer in the way of calamities for 5,000 years. These time capsules contained news that we want to transmit to posterity, so that some day those who strive to illuminate the darkness surrounding the past of their forefathers will know how we lived. The capsules are made of a metal that is harder than steel; they can survive even an atomic explosion. In addition to daily news, the capsules contain photographs of cities, ships, automobiles, aircraft, and rockets; they house samples of metals and plastics, of fabrics, threads, and cloths; they hand down to posterity objects in everyday use such as coins, tools, and toilet articles; books about mathematics, medicine, physics, biology, and astronautics are preserved on microfilm. In order to complete this service for some remote and unknown future race, the capsules also contain a "key," a book with the help of which all the written material can be translated into the languages of the future.

A group of engineers from Westinghouse Electric had the idea of presenting the time capsules to posterity. John Harrington invented the ingenious decoding system for generations yet unknown. Lunatics? Visionaries? I find the realization of this project beneficial and reassuring. It's nice to know that there are men today who think 5,000 years ahead! The archaeologists of some remote future age will not find things any easier than we did. For after an atomic conflagration none of the world's libraries will be of any use, and all the achievements that make us so proud will not be worth twopence—because they have disappeared, because they have been destroyed, because they have been atomized. An atomic conflagration which ravages the earth is not required to justify the New Yorkers' imaginative action. A shifting of the earth's axis by a few degrees would cause inundations on an unprecedented and irresistible scale—in any case they would swallow up every single written word. Who is arrogant enough to assert that the sages of old could not have conceived the same sort of idea as the farsighted New Yorkers?

Undoubtedly the strategists of an A-bomb and H-bomb war will not direct their weapons against Zulu villages and harmless Eskimos. They will use them against the centers of civilization. In other words, the radioactive chaos will fall on the advanced, most highly developed peoples. Savages and primitive peoples far away from the centers of civilization will be left. They will not be able to transmit our culture or even give an account of it, because they have never taken part in it. Even intelligent men and visionaries who tried to preserve an underground library will not have been able to help the future a great deal. "Normal" libraries will be destroyed in any case, and the surviving primitive peoples will know nothing of the hidden secret libraries. Whole regions of the globe will become burning deserts, because radiation lasting for centuries will not allow any plants to grow. The survivors will presumably be mutated, and after 2,000 years nothing will be left of the annihilated cities. The unbridled power of nature will eat its way through the ruins; iron and steel will rust and crumble into dust.

And everything will begin again! Man may embark on his adventure a second or even a third time. Perhaps once again he will take so long to reemerge as a civilized being that the secrets of old traditions and texts will be closed to him. Five thousand years after the catastrophe, archaeologists could claim that twentieth-century man was not yet familiar with iron, because, understandably enough, they would not find any, no matter how hard they dug. Along the Russian frontiers they would find miles of concrete tank traps, and they would explain that such finds undoubtedly indicated astronomical lines. If they were to find cassettes with tapes, they would not know what to do with them; they would not even be able to distinguish between played and unplayed tapes. And perhaps those tapes might hold the solution to many, many puzzles! Texts which spoke of gigantic cities with houses several hundred feet high would be pooh-poohed, because such cities could not have existed. Scholars would take the London Tube tunnels for a geometrical curiosity or an astonishingly well-conceived drainage system. And they might keep on coming across reports which described how men flew from continent to continent with giant birds and referred to extraordinary fire-spitting ships which disappeared into the sky. That would also be dismissed as mythology, because such great birds and fire-spitting ships could not have existed.

Things would be made very difficult for the translators in the year 7000. The facts about a world war in the twentieth century that they would discover from fragmentary texts would sound quite incredible. But when the speeches of Marx and Lenin fell into their hands, they would at last be able to make two high priests of this incomprehensible age the center of a religion. What a piece of luck!

People would be able to explain a great deal, provided sufficient clues were still in existence. Five thousand years is a long time. It is pure caprice on nature's part that she allows dressed blocks of stone to survive for 5,000 years. She does not deal so carefully with the thickest iron girders.

In the courtyard of a temple in Delhi there exists, as I have already mentioned, a column made of welded iron parts that has been exposed to weathering for more than 4,000 years without showing a trace of rust. In addition it is unaffected by sulphur or phosphorus. Here we have an unknown alloy from antiquity staring us in the face. Perhaps the column was cast by a group of farsighted engineers who did not have the resources for a colossal building but wanted to bequeath to posterity a visible, time-defying monument to their culture.

It is an embarrassing story: in advanced cultures of the past we find buildings that we cannot copy today with the most modern technical means. These stone masses are there; they cannot be argued away. Because that which ought not to exist cannot exist, there is a frantic search for "rational" explanations. Let us take off our blinkers and join the search. . . .

7

Ancient Marvels or Space Travel Centers?

TO the north of Damascus lies the terrace of Baalbek—a platform built of stone blocks, some of which have sides more than 65 feet long and weigh nearly 2,000 tons. Until now archaeologists have not been able to give a convincing explanation why, how, and by whom the terrace of Baalbek was built. However, Russian Professor Agrest considers it possible that the terrace is the remains of a gigantic airfield.

If we meekly accept the neat package of knowledge that the Egyptologists serve up to us, ancient Egypt appears suddenly and without transition with a fantastic ready-made civilization. Great cities and enormous temples, colossal statues with tremendous expressive power, splendid streets flanked by magnificent sculptures, perfect drainage systems, luxurious tombs carved out of the rock, pyramids of overwhelming size—these and many other wonderful things shot out of the ground, so to speak. Genuine miracles in a country that is suddenly capable of such achievements without recognizable prehistory!

Fertile agricultural land exists only in the Nile Delta and on small strips to the left and right of the river. Yet experts now estimate the number of inhabitants at the time of the building of the Great Pyramid at 50,000,000. (A figure, incidentally, that flagrantly contradicts the 20,000,000 considered to be the total population of the world in 3000 B.C.!)

With such enormous estimates a couple of million men more or less does not matter. But one thing is clear—they all had to be fed. There were not only a host of construction workers, stone masons, engineers, and sailors, there were not only hundreds of thousands of slaves, but also a well-equipped army, a large and pampered priesthood, countless merchants, farmers, and officials, and last but not least the Pharaonic household living on the fat of the land. Could they *all* have lived on the scanty yields of agriculture in the Nile Delta?

I shall be told that the stone blocks used for building the temple were moved on rollers. In other words, wooden rollers! But the Egyptians could scarcely have felled and turned into rollers the few trees, mainly palms, that then (as now) grew in Egypt, because the dates from the palms were urgently needed for food and the trunks and fronds were the only things giving shade to the dried-up ground. But they must have been wooden rollers, otherwise there would not be even the feeblest technical explanation of the building of the pyramids. Did the Egyptians import wood? In order to import wood there must have been a sizable fleet, and even after it had been landed in Alexandria the wood would have had to be transported up the Nile to Cairo. Since the Egyptians did not have horses and carts at the time of the building of the Great Pyramid, there was no other possibility. The horse-and-cart was not introduced until the seventeenth dynasty, about 1600 B.C. My kingdom for a convincing explanation of the transport of the stone blocks! Of course, the scholars say that wooden rollers were needed. . . .

There are many problems connected with the technology of the pyramid builders and no genuine solutions.

How did the Egyptians carve tombs out of the rock? What resources did they have in order to lay out a maze of galleries and rooms? The walls are smooth and mostly decorated with paintings in relief. The shafts slope down into the rocky soil; they have steps built in the best tradition of craftsmanship that lead to the burial chambers far below. Hordes of tourists stand gaping in amazement at them, but none of them gets an explanation of the mysterious technique used in their excavation. Yet it is firmly established that the Egyptians were masters of the art of tunneling from the earliest times, for the old rock-cut tombs are worked in exactly the same way as the more recent ones. There is no difference between the tomb of Tety from the sixth dynasty and the tomb of Rameses I from the New Kingdom, although there is a minimum of 1,000 years between the building of the two tombs. Obviously the Egyptians had not learned anything new to add to their old technique. In fact the more recent edifices tend increasingly to be poor copies of their ancient models.

The tourist who bumps his way to the pyramid of Cheops to the west of Cairo on a camel called Wellington or Napoleon, depending on his nationality, gets the strange sensation in the pit of his stomach that relics of the mysterious past always produce. The guide tells him that a pharaoh had a burial place built here. And with that bit of rehashed erudition he rides homeward, after taking some impressive photographs. The pyramid of Cheops, in particular, has inspired hundreds of crazy and untenable theories. In the 600-page book *Our Inheritance in the Great Pyramid,* by Charles Piazzi Smith, published in 1864, we can read about many hair-raising links between the pyramid and our globe.

Yet even after a highly critical examination, it still contains some facts that should stimulate us to reflection.

It is well known that the ancient Egyptians practiced a solar religion. Their sun god, Ra, traveled through the heavens in a bark. Pyramid texts of the Old Kingdom even describe heavenly journeys by the king, obviously made with the help of the gods and their boats. So the gods and kings of the Egyptians were also involved with flying. . . .

Is it really a coincidence that the height of the pyramid of Cheops multiplied by a thousand million—98,000,000 miles—corresponds approximately to the distance between the earth and sun? Is it a coincidence that a meridian running through the pyramids divides continents and oceans into two exactly equal halves? Is it coincidence that the area of the base of the pyramid divided by twice its height gives the celebrated figure $\pi = 3.14159$? Is it coincidence that calculations of the weight of the earth were found and is it also coincidence that the rocky ground on which the structure stands is carefully and accurately leveled?

There is not a single clue to explain why the builder of the pyramid of Cheops, the Pharaoh Khufu, chose that particular rocky terrain in the desert as the site for his edifice. It is conceivable that there was a natural cleft in the rock which he made use of for the colossal building, while another explanation, though only a feeble one, may be that he wanted to watch the progress of the work from his summer palace. Both reasons are against all common sense. In the first case it would certainly have been more practical to locate the building site nearer the eastern quarries in order to shorten transport distances, and second, it is hard to imagine that the pharaoh wanted to be disturbed year after year by the din that filled building sites day and night even in those days. Since there is so much to be said against the textbook explanations of the choice of site, one might reasonably ask whether the "gods" did not have their say here, too, even if it was by way of the priesthood. But if that explanation is accepted, there is one more important proof of my theory of the Utopian past of mankind. For the pyramid not only divides continents and oceans into two equal halves; it also lies at the center of gravity of the continents. If the facts noted here are not coincidences—and it seems extremely difficult to believe that they are—then the building site was chosen by beings who knew all about the spherical shape of the earth and the distribution of continents and seas. In this connection let us not forget Piri Reis' maps! It cannot all be coincidence or be explained away as fairy stories.

With what power, with what "machines," with what technical resources was the rocky terrain leveled at all? How did the master builders drive the tunnels downward? And how did they illuminate them? Neither here nor in the rock-cut tombs in the Valley of Kings were torches or anything similar used. There are no blackened ceilings or walls or even the slightest evidence that traces of blackening have been removed. How and with what were the stone blocks cut out of the quarries? With sharp edges and smooth sides? How were they transported and joined together to the thousandth of an inch? Once again there is a wealth of explanations for anyone to choose from: inclined planes and tracks along which the stones were pushed, scaffolding and ramps. And naturally the labor of many hundreds of thousands of Egyptian slaves: fellahin, builders, and craftsmen.

None of these explanations stands up to a critical examination. The Great Pyramid is (and remains?) visible testimony of a technique that has never been understood. Today, in the twentieth century, no architect could build a copy of the pyramid of Cheops, even if the technical resources of every continent were at his disposal.

2,600,000 gigantic blocks were cut out of the quarries, dressed and transported, and fitted together on the building site to the nearest thousandth of an inch. And deep down inside, in the galleries, the walls were painted in colors.

The site of the pyramid was a whim of the pharaoh.

The unparalleled, "classical" dimensions of the pyramid occurred to the master builder by chance.

Several hundred thousand workers pushed and pulled blocks weighing twelve tons up a ramp with (nonexistent) ropes on (nonexistent) rollers.

This host of workers lived on (nonexistent) grain.

They slept in (nonexistent) huts which the pharaoh had built outside his summer palace.

The workers were urged on by an encouraging "Heave-ho" over a (nonexistent) loudspeaker, and so the twelve-ton blocks were pushed skyward.

If the industrious workers had achieved the extraordinary daily piece rate of ten blocks piled on top of each other, they would have assembled the 2,600,000 stone blocks into the magnificent stone pyramid in about 250,000 days—664 years. Yes, and don't forget that the whole thing came into being at the whim of an eccentric king who never lived to see the completion of the edifice he had inspired.

Of course one must not even suggest that this theory, so seriously advanced, is ridiculous. Yet who is so ingenuous as to believe that the pyramid was nothing but the tomb of a king? From now on, who will consider the transmission of mathematical and astronomical signs as pure chance?

Today the Great Pyramid is undisputedly attributed to the Pharaoh Khufu as inspirer and builder. Why? Because all the inscriptions and tablets refer to Khufu. It seems obvious to me that the pyramid cannot have been erected during a single lifetime. But what if Khufu forged the inscriptions and tablets that are supposed to proclaim his fame? That was quite a popular procedure in antiquity, as many buildings bear witness. Whenever a dictatorial ruler wanted the fame for himself alone, he gave orders for this process to be carried out. If that *was* the case, then the pyramid existed long before Khufu left his visiting card.

In the Bodleian Library at Oxford there is a manuscript in which the Coptic author Mas-Udi asserts that the Egyptian King Surid had the Great Pyramid built. Oddly enough, this Surid ruled in Egypt before the Flood. And this wise King Surid ordered his priests to write down the sum total of their wisdom and conceal the writings inside the pyramid. So, according to the Coptic tradition, the pyramid was built before the Flood.

Herodotus confirms such a supposition in Book II of his *History*. The priests of Thebes had shown him 341 colossal statues, each of which stood for a high-priestly generation over a period of 11,340 years. Now we know that every high priest had his statue made during his own lifetime; and Herodotus also tells us that during his stay in Thebes one priest after another showed him his statue as a proof that the son had always followed the father. The priests assured Herodotus that their statements were very accurate, because they had written everything down for many generations, and they explained that every one of these 341 statues represented a generation. Before these 341 generations the gods had lived among men, and since then no god had visited them again in human form.

The historical period of Egypt is usually estimated at about 6,500 years. Then why did the priests lie so shamelessly to the traveler Herodotus about their 11,340 years? And why did they so expressly emphasize that no gods had dwelt among them for 341 generations? These precise details would have been completely pointless if "gods" had not really lived among men in the remote past!

We know next to nothing about the how, why, and when of the building of the pyramid. An artificial mountain, some 490 feet high and weighing 6,500,000 tons, stands there as evidence of an incredible achievement, and this monument is supposed to be nothing more than the burial place of an extravagant king! Anyone who can believe that explanation is welcome to it. . . .

Mummies, equally incomprehensible and not yet convincingly explained, stare at us from the remote past as if they held some magic secret. Various peoples knew the technique of embalming corpses, and archaeological finds favor the supposition that prehistoric beings believed in corporeal return to a second life. That interpretation would be acceptable if there was even the remotest evidence of a belief in a corporeal return in the religious philosophy of antiquity! If our primitive ancestors had believed in only a spiritual return, they would scarcely have gone to such trouble with the dead. But finds in Egyptian tombs provide example after example of the preparation of embalmed corpses for a corporeal return.

What the evidence says, what visible proof says, cannot be so absurd! Drawings and sagas actually indicated that the "gods" promised to return from the stars in order to awaken the well-preserved bodies to new life. That is why the provisioning of the embalmed corpses in the burial chambers took such a practical form and was intended for a life on this side of the grave. Otherwise what were they supposed to have done with money, jewelry, and their favorite articles? And as they were even provided in the tomb with some of their servants, who were unquestionably buried alive, the point of all the preparations was obviously the continuation of the old life in a new life. The tombs were tremendously durable and solid, almost atom-bomb-proof; they could survive the ravages of all the ages. The valuables left in them, gold and precious stones, were virtually indestructible. I am not concerned here with discussing the later abuses of mummification. I am only concerned with the question: Who put the idea of corporeal rebirth into the heads of the heathen? And whence came the first audacious idea that the cells of the body had to be preserved so that the corpse, preserved in a very secure place, could be awakened to new life after thousands of years?

So far this mysterious reawakening complex has only been considered from the religious point of view. But supposing the pharaoh, who certainly knew more about the nature and customs of the "gods" than his subjects, had these possibly quite crazy ideas? "I must make a burial place for myself that cannot be destroyed for millennia and is visible far across the country. The gods promised to return and wake me up (or doctors in the distant future will discover a way to restore me to life again)."

What have we to say about that in the space age?

In his book *The Prospect of Immortality,* published in 1965, physican and astronomer Robert C. W. Ettinger suggests a way in which twentieth-century men can have ourselves frozen so that our cells can go on living from the medical and biological point of view, but slowed down a billionfold. For the present this idea may still sound Utopian, but in fact every big clinic today has a "bone bank" which preserves human bones in a deep-frozen condition for years and makes them serviceable again when required. Fresh blood—this too is a universal practice—can be kept for an unlimited time at minus 196° C, and living cells can be stored almost indefinitely at the temperature of liquid nitrogen. Did the pharaoh have a fantastic idea which will soon be realized in practice?

You must read what follows twice to grasp the fantastic implications of the result of the next piece of scientific research. In March, 1963, biologists of the University of Oklahoma confirmed that the skin cells of the Egyptian Princess Mene were capable of living. And Princess Mene has been dead for several thousand years!

There have been finds in many places of mummies which are preserved so completely and intact that they seem to be alive. Glacier mummies left by the Incas survived the ages and theoretically they are capable of living. Utopia? In the summer of 1965 Russian television showed two dogs which had been deep-frozen for a week. On the seventh day they were thawed out again and—hey, presto!—they went on living as cheerfully as ever!

Americans—and this is no secret—are seriously concerned, as part of their space program, with the problem of how to freeze astronauts of the future for their long journeys to distant stars.

Dr. Ettinger, often scoffed at today, prophesies a remote future in which men will not be consumed by fire or eaten by worms—a future in which bodies, frozen in deep-freeze cemeteries or deep-freeze bunkers, await the day when advances in medical science can remove the cause of their death and bring their bodies to new life. One can see the terrifying vision of an army of deep-frozen soldiers who will be thawed out as necessary in case of war. A really horrifying idea.

But what connection have mummies with our theory of space travelers in the remote past? Am I dragging proofs in willy-nilly?

I ask: How did the ancients know that the body cells continue to live slowed down a billionfold after special treatment?

I ask: Where did the idea of immortality come from, and how did people get the concept of corporeal reawakening in the first place?

The majority of ancient peoples knew the technique of mummification, and the rich people actually practiced it. I am not concerned here with this demonstrable fact, but with solving the problem of where the idea of a reawakening, a return to life, originated. Did the idea occur to some king or tribal prince purely by chance or did some prosperous citizen watch "gods" treating their corpses with a complicated process and preserving them in bomb-proof sarcophagi? Or did some "gods" (space travelers) transmit to a quick-witted prince of royal blood their knowledge of how corpses can be reawakened after a special treatment?

These speculations require confirmation from contemporary sources. In a few hundred years mankind will have a mastery of space travel that is inconceivable today. Travel agencies will offer trips to the planets, with precise departure and return dates, in their brochures. Obviously a prerequisite for this mastery is that all branches of science keep pace with the development of space travel. Electronics and cybernetics alone will not do the trick. Medicine and biology will make their contribution by finding out ways of lengthening the vital functions of human beings. Today this department of space research is also working in top gear. Here we must ask ourselves: Did space travelers in prehistory already possess knowledge that we must gain anew? Did unknown intelligences already know the methods with which to treat bodies so that they could be revived in so many thousand years? Perhaps the "gods," being shrewd, had an interest in "preserving" at least one dead man with all the knowledge of his time so that some day he could be questioned about the history of his generation? Who can tell? Is it not possible that such an interrogation by "gods" who came back has already taken place?

In the course of the centuries, mummification, originally a solemn matter, became the fashion. Suddenly everyone wanted to be reawakened; suddenly everyone thought that he would come to new life so long as he did the same as his forefathers. The high priests, who actually did possess some knowledge of such reawakenings, did a great deal to encourage this cult, for their class did good business out of it.

I have already mentioned the physically impossible ages of the Sumerian kings and the biblical figures. I asked whether these people could not have been space travelers who prolonged their life-span through the effect of the time shift on interstellar flights just below the speed of light.

Are we perhaps getting a clue to the incredible age of the men named in the texts if we assume that they were mummified or frozen? If we follow this theory, then the unknown space travelers would have frozen leading personalities in antiquity—put them into an artificial deep sleep, as legends tell us—and taken them out of the drawer, thawed them out, and conversed with them during subsequent visits. At the end of each visit it would have been the task of the priestly class appointed and instructed by the space travelers to prepare the living dead again and preserve them once more in giant temples until the "gods" returned.

Impossible? Ridiculous? It is mostly those people who feel that they are absolutely bound by the laws of nature who make the most stupid objections. Does not nature herself display brilliant examples of "hibernation" and reawakening?

There are species of fish which, after being frozen stiff, thaw out at milder temperatures and swim around again in the water. Flowers, larvae, and grubs not only go into hibernation but also reappear in the spring in lovely new garb.

Let me be my own devil's advocate. Did the Egyptians learn the possibility of mummification from nature? If that were the case, there ought to have been a cult of butterflies or cockchafers or at least a trace of such a cult. There is nothing of the kind. Underground tombs do contain gigantic sarcophagi with mummified animals, but given their climate the Egyptians could not have copied hibernation from animals.

Five miles from Helwan lie more than 5,000 tombs of different sizes which all date to the time of the first and second dynasties. These tombs show that the art of mummification is more than 6,000 years old.

In 1953 Professor Emery discovered a large tomb in the archaic cemetery of North Sakkara that is attributed to a pharaoh of the first dynasty. Apart from the main tomb there were 72 other tombs, arranged in three rows, in which lay the bodies of the servants who wanted to accompany their king in the new world. No trace of violence is visible on the bodies of the 64 young men and 8 young women. Why did these 72 allow themselves to be walled up and killed?

Belief in a second life beyond the grave is the best-known

and also the simplest explanation of this phenomenon. In addition to gold and jewelry the pharaoh was provided in the tomb with grain, oil, and spices which were obviously intended as provisions for the life to come. Apart from grave robbers, the tombs were also opened by later pharaohs. In such cases the pharaoh found the provisions in the tomb of his ancestor well preserved. In other words the dead man had neither eaten them nor taken them into another world. And when the tomb was closed again, fresh supplies were placed in the vault, which was shut up, protected against thieves, and sealed with many traps. It seems obvious that the Egyptians believed in a reawakening in the distant future, not an immediate reawakening in the hereafter.

In June, 1954, also at Sakkara, a tomb was discovered that had not been robbed, for a chest containing jewels and gold lay in the burial chamber. The sarcophagus was closed with a sliding lid, instead of a removable one. On June 9 Dr. Goneim ceremonially opened the sarcophagus. It contained nothing. Absolutely nothing. Did the mummy decamp, leaving its jewels behind?

The Russian Rodenko discovered a grave, Kurgan V, fifty miles from the frontier of Outer Mongolia. This grave takes the form of a rocky hill that is faced internally with wood. All the burial chambers are packed with eternal ice, and as a result the contents of the grave were preserved in a state of deep-freeze. One of these chambers contained an embalmed man and a similarly treated woman. Both of them were provided with everything that they might have needed for a life to come: foodstuffs in dishes, clothes, jewels, and musical instruments. Everything was deep-frozen and in an excellent state of preservation, including the naked mummies. In one burial chamber scholars identified a rectangle containing four rows of six squares, each of which had a drawing inside it. The whole could be a copy of the stone carpet in the Assyrian palace at Nineveh! Strange sphinxlike figures with complicated horns on their heads and wings on their backs are clearly visible, and their posture shows them to be aspiring skyward.

But motives for a second spiritual life can scarcely be based on the finds in Mongolia. The deep-freezing used in the graves there—for that is what the chambers faced with wood and filled with ice amount to—is too much of this world and obviously intended for terrestrial ends. Why, and this question keeps on worrying us, did the ancients think that bodies prepared in this way achieved a state which would make reawakening possible? That is a puzzle for the time being.

In the Chinese village of Wu Chuan is a rectangular tomb measuring 45 by 39 feet; in it lie the skeletons of 17 men and 24 women. Here, too, none of the skeletons shows signs of a violent death. There are glacier tombs in the Andes, ice tombs in Siberia, group and individual graves in China, Sumeria, and Egypt. Mummies have been found in the far north and in South Africa. And all the dead were supplied with the necessities for a new life, and all the tombs were so planned and built that they could survive for thousands of years.

Is it all mere coincidence? Are they all merely individual fancies, strange whims on the part of our ancestors? Or is there an ancient promise of corporeal return that is unknown to us? Who could have made it?

Some 10,000-year-old tombs were excavated at Jericho, and a number of 8,000-year-old heads, modeled in plaster of Paris, were found. That, too, is astonishing, for ostensibly this people did not know the techniques of pottery making. In another part of Jericho whole rows of round houses were discovered. The walls are curved inward at the top, like domes.

The omnipotent carbon isotope C-14, with the aid of which the age of organic substances can be determined, gives dates with a maximum of 10,400 years in this case. These scientifically determined dates agree fairly well with the dates which the Egyptian priests transmitted. They said that their priestly ancestors had discharged their duties for more than 11,000 years. Is this only a coincidence, too?

Prehistoric stones at Lussac, France, form a particularly remarkable find. They show drawings of men dressed in completely modern style, with hats, jackets, and short trousers. Abbé Breuil says that the drawings are authentic, and his statement throws the whole of prehistory into confusion. Who engraved the stones? Who has enough imagination to conceive of a caveman dressed in skins who drew figures from the twentieth century on the walls?

Some really magnificent Stone Age paintings were found in 1940 in the Lascaux caves in the South of France. The paintings in this gallery are as lively and intact as if they had been done today, and two questions immediately spring to mind. How was this cave illuminated for the laborious work of the Stone Age artists, and why were the walls decorated with these astonishing paintings?

Let the people who consider these questions stupid explain the contradictions. If the Stone Age cavemen were primitive and savage, they could not have produced the astounding paintings on the cave walls. But if the savages were capable of painting these pictures, why should they not also have been able to build huts as shelter? The foremost authorities concede that animals had the ability to build nests and shelters millions of years ago. But it obviously does not fit into the working hypothesis to concede *homo sapiens* the same ability as long ago as that.

In the Gobi Desert, deep down below the ruins of Khara Khota—not far from those strange sand vitrifications which can only have taken place under the influence of tremendous heat—Professor Koslov found a tomb that is dated to about 12,000 years B.C. A sarcophagus contained the bodies of two rich men, and the sign of a circle bisected vertically was found on the sarcophagus.

In the Subis Mountains on the west coast of Borneo a

network of caves was found that had been hollowed out on a cathedral-like scale. Among these colossal finds there are fabrics of such fineness and delicacy that with the best will in the world one cannot imagine savages making them. Questions, questions, questions. . . .

The first doubts are beginning to insinuate themselves into sterotyped archaeological theory, but what we need to do is to force breaches in the thicket of the past. Landmarks must be set up again; wherever possible, a new series of fixed dates must be established.

Let me make it clear that I am not doubting the history of the last 2,000 years here. I am speaking solely and exclusively of the most remote antiquity, of the blackest darkness of time, which I am striving to illuminate by asking new questions.

Nor can I give any figures and dates showing when the visit of unknown intelligences from the universe began to influence our young intelligences. But I venture to doubt the current datings applied to the remote past. I would suggest, on tolerably good grounds, placing the incident I am concerned with in the Early Paleolithic Age—between 10,000 and 40,000 B.C. Our hitherto existing methods of dating, including C-14, which makes everyone so happy, leave great gaps as soon as we have to deal with periods of fewer than 5,000 years. The older the substance to be examined, the more unreliable the radiocarbon method is. Even recognized scholars have told me that they consider the C-14 method to be an out-and-out bluff, because if an organic substance is from 30,000 to 50,000 years old its age can be established anywhere between those limits.

These critical voices should be accepted only with limitations; nevertheless, a second dating method parallel to the C-14 method and based on the latest measuring apparatus would unquestionably be desirable.

8

Easter Island—Land of the Bird Men

THE first European seafarers who landed on Easter Island at the beginning of the eighteenth century could scarcely believe their eyes. On this little plot of earth, 2,350 miles from the coast of Chile, they saw hundreds of colossal statues lying scattered about all over the island. Whole mountain massifs had been transformed, steel-hard volcanic rock had been cut through like butter, and 10,000 tons of massive rocks lay in places where they could not have been dressed. Hundreds of gigantic statues, some of which are between 33 and 66 feet high and weigh as much as 50 tons, still stare challengingly at the visitor today—like robots which seem to be waiting solely to be set in motion again. Originally these colossuses also wore hats; but even the hats do not exactly help to explain the puzzling origin of the statues. The stone for the hats, which weighed more than ten tons apiece, was found at a different site from that used for the bodies, and in addition the hats had to be hoisted high in the air.

Wooden tablets, covered with strange hieroglyphs, were also found on some of the statues in those days. But today it is impossible to find more than ten fragments of those tablets in all the museums in the world, and none of the inscriptions on those still extant has yet been deciphered.

Thor Heyerdahl's investigations of these mysterious giants produced three clearly distinguishable cultural periods, and the oldest of the three seems to have been the most perfect. Heyerdahl dates some charcoal remains that he found to about A.D. 400. It has not been proved whether the fireplaces and remains of bones had any connection with the stone colossuses. Heyerdahl discovered hundreds of unfinished statues near rock faces and on the edges of craters; thousands of stone implements, simple stone axes, lay around as if the work had been abandoned quite suddenly.

Easter Island lies far away from any continent or civilization. The islanders are more familiar with the moon and the stars than any other country. No trees grow on the island, which is a tiny speck of volcanic stone. The usual explanation, that the stone giants were moved to their present sites on wooden rollers, is not feasible in this case, either. In addition, the island can scarcely have provided food for more than 2,000 inhabitants. (A few hundred natives live on Easter Island today.) A shipping trade, which brought food and clothing to the island for the stonemasons, is hardly credible in antiquity. Then who cut the statues out of the rock, who carved them and transported them to their sites? How were they moved across country for miles without rollers? How were they dressed, polished, and erected? How were the hats, the stone for which came from a different quarry from that of the statues, put in place?

Even if people with lively imaginations have tried to picture the Egyptian pyramids being built by a vast army of workers using the "heave-ho" method, a similar method would have been impossible on Easter Island for lack of manpower. Even 2,000 men, working day and night, would not be nearly enough to carve these colossal figures out of the steel-hard volcanic stone with rudimentary tools—and at least a part of the population must have tilled the barren fields, gone fishing, woven cloth, and made ropes. No, 2,000 men alone could not have made the gigantic statues. And a larger population is inconceivable on Easter Island. Then who did do the work? And how did they manage it? And why do the statues stand around the edge of the island and not in the interior? What cult did they serve?

Unfortunately, the first European missionaries on this tiny patch of earth helped to ensure that the island's dark ages stayed dark. They burned the tablets with hieroglyphic characters; they prohibited the ancient cults of the gods and did away with every kind of tradition. Yet thoroughly as the pious gentlemen went to work, they could not prevent the natives from calling their island the Land of the Bird Men, as they still do today. An orally transmitted legend tells us that flying men landed and lighted fires in ancient times. The legend is confirmed by sculptures of flying creatures with big, staring eyes.

Connections between Easter Island and Tiahuanaco automatically force themselves upon us. There as here, we find stone giants belonging to the same style. The haughty faces with their stoic expressions suit the statues—here as there. When Francisco Pizarro questioned the Incas about Tiahuanaco in 1532, they told him that no man had ever seen the city save in ruins, for Tiahuanaco had been built in the night of mankind. Traditions call Easter Island the "navel of the world." It is more than 3,125 miles from Tiahuanaco to Easter Island. How can one culture possibly have inspired the other?

Perhaps pre-Inca mythology can give us a hint here. In it the old god of creation, Viracocha, was an ancient and elemental divinity. According to tradition Viracocha created the world when it was still dark and had no sun; he sculpted a race of giants from stone, and when they displeased him, he sank them in a deep flood. Then he caused the sun and the moon to rise above Lake Titicaca, so that there was light on earth. Yes, and then—read this closely—he shaped clay figures of men and animals at Tiahuanaco and breathed life into them. Afterward, he instructed these living creatures of his own creation in language, customs, and arts, and finally flew some of them to different continents which they were supposed to inhabit thenceforth. After this task the god Viracocha and two assistants traveled to many countries to check how his instructions were being followed and what results they had had. Dressed as an old man, Viracocha wandered over the Andes and along the coast, and often he was given a poor reception. Once, at Cacha, he was so annoyed by his welcome that in a fury he set fire to a cliff which began to burn up the whole country. Then the ungrateful people asked his forgiveness, whereupon he extinguished the flames with a single gesture. Viracocha traveled on, giving instructions and advice, and many temples were erected to him as a result. Finally he said good-bye in the coastal province of Manta and disappeared over the ocean, riding on the waves, but he said he intended to come back.

The Spanish conquistadors who conquered South and Central America came up against the sagas of Viracocha everywhere. Never before had they heard of gigantic white men who came from somewhere in the sky. Full of astonishment, they learned about a race of sons of the sun who instructed mankind in all kinds of arts and disappeared again. And in all the legends that the Spaniards heard, there was an assurance that the sons of the sun would return.

Although the American continent is the home of ancient cultures, our accurate knowledge of America is barely 1,000 years old. It is an absolute mystery to us why the Incas cultivated cotton in Peru in 3000 B.C., although they did not know or possess the loom. The Mayas built roads but did not use the wheel, although they knew about it. The fantastic five-strand necklace of green jade in the burial pyramid of

Tikal in Guatemala is a miracle. A miracle because the jade comes from China. The sculptures of the Olmecs are incredible. With their beautifully helmeted giant skulls, they can be admired only on the sites where they were found, for they will never be on show in a museum. No bridge in the country could stand their weight. We can move smaller "monoliths" weighing up to fifty tons with our modern lifting appliances and loaders, but when it comes to hundred-tonners like these our technology breaks down. But our ancestors could transport and dress them. How?

It even seems as if the ancient peoples took a special pleasure in juggling with stone giants over hill and dale. The Egyptians fetched their obelisk from Aswan, the architects of Stonehenge brought their stone blocks from southwest Wales and Marlborough, the stonemasons of Easter Island took their ready-made monster statues from a distant quarry to their present sites, and no one can say where some of the monoliths at Tiahuanaco come from. Our remote ancestors must have been strange people; they liked making things difficult for themselves and always built their statues in the most impossible places. Was it just because they liked a hard life?

I refuse to think that the artists of our great past were as stupid as that. They could just as easily have erected their statues and temples in the immediate vicinity of the quarries if an old tradition had not laid down where their works ought to be sited. I am convinced that the Inca fortress of Sacsahuamán was not built above Cuzco by chance, but rather because a tradition indicated the place as a holy spot. I am also convinced that in all the places where the most ancient monumental buildings of mankind were found the most interesting and important relics of our past lie still untouched in the ground, relics, moreover, which could be of tremendous importance for the further development of present-day space travel.

The unknown space travelers who visited our planet many thousands of years ago can hardly have been less farsighted than we think we are today. They were convinced that one day man would make the move out into the universe on his own initiative, using his own skills.

It is a well-known historical fact that the intelligences of our planet have constantly sought for kindred spirits, for life, for corresponding intelligences in the cosmos.

Present-day antennae and transmitters have broadcast the first radio impulses to unknown intelligences. When we shall receive an answer—in ten, fifteen, or a hundred years —we do not know. We do not even know which star we should beam our message at, because we have no idea which planet should interest us most. Where do our signals reach unknown intelligences similar to human beings? We do not know. Yet there is much to support the belief that the information needed to reach our goal is deposited in our earth for us. We are trying hard to neutralize the force of gravity; we are experimenting with elementary particles and antimatter. Are we also doing enough to find the data which are hidden in our earth, so that we can at last ascertain our original home?

If we take things literally, much that was once fitted into the mosaic of our past with great difficulty becomes quite plausible: not only the relevant clues in ancient texts but also the "hard facts" which offer themselves to our critical gaze all over the globe. Lastly, we have our reason to think with.

So it will be man's ultimate insight to realize that his justification for existence to date and all his struggles to advance really consisted in learning from the past in order to make himself ready for contact with the existence in space. Once that happens, the shrewdest, most die-hard individualist must see that the whole human task consists in colonizing the universe and that man's whole spiritual duty lies in perpetuating all his efforts and practical experience. Then the promise of the "gods" that peace will come on earth and that the way to heaven is open can come true.

As soon as the available authorities, powers, and intellects are devoted to space research, the results will make the absurdity of terrestrial wars abundantly clear. When men of all races, peoples, and nations unite in the supranational task of making journeys to distant planets technically feasible, the earth with all its mini-problems will fall back into its right relation with the cosmic processes.

Occultists can put out their lamps, alchemists destroy their crucibles, secret brotherhoods take off their cowls. It will no longer be possible to offer man the nonsense that has been purveyed to him so brilliantly for thousands of years. Once the universe opens its doors, we shall attain a better future.

I base the reasons for my skepticism about the interpretation of our remote past on the knowledge that is available today. If I admit to being a skeptic, I mean the word in the sense in which Thomas Mann used it in a lecture in the twenties: "The positive thing about the skeptic is that he considers everything possible!"

9

The Mysteries of South America and Other Oddities

ALTHOUGH I emphasized that it is not my intention to call in question the history of mankind during the last 2,000 years, I believe that the Greek and Roman gods and also most of the figures in sagas and legends are surrounded by the breath of a very remote past. Since mankind has existed, age-old traditions have lived on among the various peoples. More recent cultures, too, provide us with indications pointing to the remote, unknown past.

Ruins in the jungles of Guatemala and Yucatán can bear comparison with the colossal edifices of Egypt. The ground area of the pyramid of Cholula, 60 miles south of the Mexican capital, is bigger than that of the pyramid of Cheops. The pyramid field of Teotihuacán, 25 miles north of Mexico City, covers an area of almost 8 square miles, and all the edifices are aligned according to the stars. The oldest text about Teotihuacán tells us that the gods assembled here and took council about man, even before *homo sapiens* existed!

The calendar of the Mayas, the most accurate in the world, has already been mentioned, and so has the Venusian formula. Today it is proved that all the edifices at Chichén Itzá, Tikal, Copán, and Palenque were built according to the fabulous Mayan calendar. The Mayas did not build pyramids because they needed them; they did not build temples because they needed them; they built temples and pyramids because the calendar decreed that a fixed number of steps of a building had to be completed every 52 years. Every stone has its relation to the calendar; every completed building conforms exactly to certain astronomical requirements.

But an absolutely incredible thing happened about A.D. 600! Suddenly, and for no apparent reason, a whole people left its laboriously and solidly built cities, with their rich temples, artistic pyramids, squares lined with statues, and grandiose stadiums. The jungle ate its way through buildings and streets, broke up the masonry, and turned everything into a vast landscape of ruins. No inhabitant ever returned there.

Let us pretend that this event, this enormous national migration, happened in ancient Egypt. For generations the people built temples, pyramids, cities, water conduits, and streets according to calendar dates; wonderful sculptures were laboriously carved out of stone with primitive tools and installed in the magnificent buildings; when this work, lasting more than a millennium, was finished, they left their homes and moved to the barren north. Such a procedure, brought a little closer to the course of historical events that we are familiar with, seems incredible because it is ridiculous. The more incomprehensible a procedure, the more numerous the vague explanations and attempts at interpretation. The first version put forward was that the Mayas might have been driven out by foreign invaders. But who could have overcome the Mayas, who were at the peak of their civilization and culture? No traces that could be connected with a military confrontation have ever been found. The idea that the migration could have been caused by a marked change in climate is well worth considering. But there are no signs to support this view either. The distance covered by the

Mayas from the territory of the old to the borders of the new kingdom measures only 220 miles as the crow flies—a distance that would have been inadequate to escape a catastrophical change in climate. The explanation that a devastating epidemic set the Mayas on the move also deserves serious examination. Apart from the fact that this explanation is offered as one of many, there is not the slightest proof of it. Was there a battle between the generations? Did the young revolt against the old? Was there a civil war, a revolution? If we opt for one of these possibilities, it is obvious that only a part of the population, namely the defeated, would have left the country and that the victors would have remained in their old settlements. Investigations of archaeological sites have not produced one proof that even a single Maya remained behind. The whole people suddenly emigrated, leaving their holy places unguarded in the jungle.

I should like to introduce a new note into the concert of opinions, a theory that is not proved any more than the other interpretations are. But regardless of the probability of the other explanations, I venture to make my contribution boldly and with conviction.

At some point in a very early period the Mayas' ancestors were paid a visit by the "gods" (in whom I suspect space travelers). As a number of factors support the assumption, the ancestors of the American cultural peoples may perhaps have immigrated from the ancient Orient. But in the world of the Mayas there were strictly guarded sacred traditions about astronomy, mathematics, and the calendar! The priests guarded the traditional knowledge because the "gods" had given their word to return one day. They created a grandiose new religion, the religion of Kukulkán, the Feathered Serpent.

According to priestly tradition, the gods would come back from heaven when the vast buildings were completed according to the laws of the calendar cycle. So the people hastened to complete temples and pyramids according to this holy rhythm, because the year of completion was supposed to be a year of rejoicing. Then the god Kukulkán would come from the stars, take possession of the buildings, and from then on live among mankind.

The work was finished, the year of the god's return came around—but nothing happened. The people sang, prayed, and waited for a whole year. Slaves and jewelry, corn and oil, were offered up in vain. But heaven remained dumb and without a sign. No heavenly chariot appeared; they could hear no rushing or distant thunder. Nothing, absolutely nothing, happened.

If we give this hypothesis a chance, the disappointment of priests and people must have been tremendous. The work of centuries had been done in vain. Doubts arose. Was there a mistake in the calculation of the calendar? Had the gods landed somewhere else? Had they all made a terrible mistake?

I should mention that the mystical year of the Mayas, in which the calendar began, goes back to 3111 B.C. Proofs of this exist in Mayan writings. If we accept this date as proved, then there was only a gap of a few hundred years between it and the beginning of the Egyptian culture. This legendary age seems to be genuine, because the hyperaccurate Mayan calendar says so over and over again. If that is so, the calendar and the national migration are not the only things that make me skeptical. For a comparatively new find starts off nagging doubts, too.

In 1935 a stone relief that very probably represents the god Kukumatz (in Yucatán, Kukulkán) was found in Palenque (Old Kingdom). A genuinely unprejudiced look at this picture would make even the most die-hard skeptic stop and think.

There sits a human being, with the upper part of his body bent forward like a racing motorcyclist; today any child would identify his vehicle as a rocket. It is pointed at the front, then changes to strangely grooved indentations like inlet ports, widens out, and terminates at the tail in a darting flame. The crouching being himself is manipulating a number of indefinable controls and has the heel of his left foot on a kind of pedal. His clothing is appropriate: short trousers with a broad belt, a jacket with a modern Japanese opening at the neck, and closely fitting bands at arms and legs. With our knowledge of similar pictures, we should be surprised if the complicated headgear were missing. And there it is with the usual indentations and tubes, and something like antennae on top. Our space traveler—he is clearly depicted as one—is not only bent forward tensely; he is also looking intently at an apparatus hanging in front of his face. The astronaut's front seat is separated by struts from the rear portion of the vehicle, in which symmetrically arranged boxes, circles, points, and spirals can be seen.

What does this relief have to tell us? Nothing? Is everything that anyone links up with space travel a stupid figment of the imagination?

If the stone relief from Palenque is also rejected from the chain of proofs, one must doubt the integrity which scholars bring to the investigation of outstanding finds. After all, one is not seeing ghosts when one is analyzing actual objects.

To continue with our series of hitherto unanswered questions: Why did the Mayas build their oldest cities in the jungle, and not on a river, or by the sea? Tikal, for example, lies 109 miles as the crow flies from the Gulf of Honduras, 161 miles northwest of the Bay of Campeche, and 236 miles as the crow flies north of the Pacific Ocean. The fact that the Mayas were quite familiar with the sea is shown by the wealth of objects made of coral, mussels, and shellfish. Why, then, the "flight" into the jungle? Why did they build water reservoirs when they could have settled by the water? In Tikal alone there are 13 reservoirs with a capacity of 214,504 cubic yards.

Why did they absolutely have to live, build, and work here and not in some more "logically" situated place?

After their long trek the disappointed Mayas founded a new kingdom in the north. And once again cities, temples, and pyramids arose according to the dates prefixed by the calendar.

To give some idea of the accuracy of the Mayan calendar, here are the periods of time they used:

20 kins	= 1 uinal or 20 days
18 uinals	= 1 tun or 360 days
20 tuns	= 1 katun or 7,200 days
20 katuns	= 1 baktun or 144,000 days
20 baktuns	= 1 pictun or 2_{1},880,000 days
20 pictuns	= 1 calabtun or 56_{1},600,000 days
20 calabtuns	= 1 kinchiltun or 1_{2},152_{1},000,000 days
20 kinchiltuns	= 1 atautun or 23_{2},040_{1},000,000 days

But the stone steps based on calendar dates are not the only things that tower above the green roof of the jungle, for observatories were built, too.

The observatory at Chichén is the first and oldest round building of the Mayas. Even today the restored building looks like an observatory. The circular edifice rises far above the jungle on three terraces; inside it a spiral staircase leads to the uppermost observation post; in the dome there are hatches and openings directed at the stars and giving an impressive picture of the firmament at night. The outer walls bear masks of the rain god . . . and the image of a human figure with wings.

Admittedly, the Mayas' interest in astronomy is not sufficient motivation for our hypothesis of relations with intelligences on other planets. The abundance of hitherto unanswered questions is bewildering: How did the Mayas know about Uranus and Neptune? Why are the observation posts in the observatory at Chichén not directed at the brightest stars? What does the stone relief of the rocket-driving god at Palenque mean? What is the point of the Mayan calendar with its calculations for 400,000,000 years? Where did they get the knowledge required to calculate the solar and Venusian years to four decimal places? Who transmitted their inconceivable astronomical knowledge? Is every fact a chance product of the Mayan intellect or does each fact, or rather do all the facts added together, conceal a revolutionary message for a very distant future, as seen from their point in time?

If we put all the facts in a sieve and roughly separate the wheat from the chaff, there are so many inconsistencies and absurdities left that research needs spurring on to make a large-scale new effort to solve at least some of the enormous number of problems. For in our age research should no longer remain satisfied when confronted with so-called "impossibilities."

I have one more, rather gruesome story to tell, the story of the sacred well of Chichén Itzá. From the stinking mud of this well Edward Herbert Thompson excavated not only jewelry and objects of art but also the skeletons of youths. Drawing on ancient accounts, Diego de Landa stated that in times of drought the priests used to make pilgrimages to the well to appease the wrath of the rain god by throwing boys and girls into it during a solemn ceremony.

Thompson's finds proved De Landa's claim. A horrifying story, which also brings up more questions from the bottom of the well. How did this water hole come into being? Why was it declared a sacred well? Why this well in particular, for there are several like it?

The exact counterpart of the sacred well of Chichén Itzá exists, hidden in the jungle, barely 76 yards from the Mayan observatory. Guarded by snakes, poisonous millipedes, and troublesome insects, the hole has the same measurements as the "real" well; its vertical walls are equally weathered, overgrown, and swamped by the jungle. These two wells resemble each other most strikingly. The water is the same depth and the color shimmers from green to brown and blood-red in both wells. Unquestionably the two wells are the same age, and possibly they both owe their existence to the impact of meteorites. Meanwhile, contemporary scholars speak only of the sacred well of Chichén Itzá; the second well, which is so similar, does not fit into their theories, although both wells are 984 yards away from the top of the biggest pyramid, the Castillo. This pyramid belongs to the god Kukulkán, the Feathered Serpent.

The snake is a symbol of nearly all Mayan buildings. That is astonishing, for one would have expected a people surrounded by luxuriant rampant flora to leave flower motifs behind on their stone reliefs as well. Yet the loathsome snake confronts us everywhere. From time immemorial the snake has wound its way through the dust and dirt of the earth. Why should anyone conceive of endowing it with the ability to fly? Primeval image of evil, the snake is condemned to crawl. How could anyone worship this repulsive creature as a god, and why could it fly as well? Among the Mayas it could. The god Kukulkán (Kukumatz) presumably corresponds to the figure of the later god Quetzalcoatl. What does the Mayan legend tell us about this Quetzalcoatl?

He came from an unknown country of the rising sun in a white robe, and he wore a beard. He taught the people all the sciences, arts, and customs, and left very wise laws. It was said that under his direction corncobs grew as big as a man and that cotton grew already colored. When Quetzalcoatl had fulfilled his mission, he returned to the sea, preaching his teaching en route, and boarded a ship there which took him to the morning star. It is almost embarrassing for me to mention in addition that the bearded Quetzalcoatl also promised to return.

Naturally there is no lack of explanations for the appearance of the wise old man. A kind of messianic role is attributed to him, for a man with a beard is not an everyday occurrence in these latitudes. There is even a daring version which suggests that the old Quetzalcoatl was an earlier Jesus! That does not convince me. Anyone who arrived among the Mayas from the ancient world would have known about the wheel for transporting men and objects. Surely one of the first actions of a sage, a god like Quetzalcoatl, who appeared as missionary, lawgiver, doctor, and adviser on many practical aspects of life, would have been to instruct the poor Mayas in the use of the wheel and the cart. In fact the Mayas never used either.

Let us complete the intellectual confusion with a compendium of oddities from the dim past.

In 1900 Greek sponge divers found an old wreck loaded with marble and bronze statues off Antikythera. These art treasures were rescued, and subsequent investigations showed that the ship must have foundered around the time of Christ. When all the plunder was sorted out, it included a shapeless lump that proved more important than all the statues put together. When it had been carefully treated, scholars discovered a sheet of bronze with circles, inscriptions, and cog wheels and soon realized that the inscriptions must be connected with astronomy. When the many separate parts were cleaned, a strange construction came to light, a regular machine with movable pointers, complicated scales or dials and metal plates with writing. The reconstructed machine had more than twenty little wheels, a kind of differential gear, and a crown wheel. On one side was a spindle that set all the dials in motion at varying speeds as soon as it was turned. The pointers were protected by bronze covers on which long inscriptions could be read. In the case of this "machine from Antikythera," is there the slightest doubt that first-class precision mechanics were at work in antiquity? Moreover, the machine is so complicated that it was probably not the first of its kind. Professor Solla Price interpreted the apparatus as a kind of calculating machine with the help of which the movements of the moon, the sun, and probably other planets could be worked out.

The fact that the machine gives the year of its construction as 82 B.C. is not so important. It would be more interesting to find out who built the first model of this machine, this small-scale planetarium!

The Hohenstaufen Emperor Frederick II is supposed to have brought back a most unusual tent from the east when he returned from the Fifth Crusade in 1229. In the interior of the tent stood a clockwork motor, and people watched the constellations in motion through the dome-shaped roof of the tent. Once again, a planetarium in olden times. We accept its existence at that date because we know that the necessary mechanical skills existed then. The idea of the earlier planetarium irritates us because in Christ's day the concept of a heaven with fixed stars taking into account the rotation of the earth did not exist. Even the knowledgeable Chinese and Arabic astronomers of antiquity can give us no help regarding this inexplicable fact, and it is undeniable that Galileo Galilei was not born until 1,500 years later. Anyone who goes to Athens should not miss the "machine from Antikythera"; it is on view in the National Archaeological Museum. We possess only written accounts of Frederick II's tent planetarium.

Here are some more strange things that antiquity has bequeathed to us:

Outline drawings of animals which simply did not exist in South America 10,000 years ago, namely camels and lions, were found on the rocks of the desert plateau of Marcahuasi 12,500 feet above sea level.

In Turkestan engineers found semicircular structures made of a kind of glass or pottery. Their origin and significance cannot be explained by the archaeologists.

The ruins of an ancient town which must have been de-

Above: The ancient Mayan Temple of the Inscriptions at Palenque, Mexico.

Left: Inside on the wall of one of the small chambers is this relief engraving. There is not enough space to take a full front-view photograph, but there is enough detail to check the artist's drawing reproduced on the next page.

Three curious things from Assyria and one from Iraq. The top tablet shows the god Shamasi. It is from the third millennium B.C. and shows stars and figures with peculiar headgear. Why should ancient gods be associated with the stars?

The second tablet is from the early first millennium B.C. The object in the center is described as a sacred tree. It could just as reasonably be interpreted as a symbolic representation of the construction of an atom, with an astronaut in a fiery chariot above. We have a figure within a winged circle and below the circle a representation of a propulsion unit.

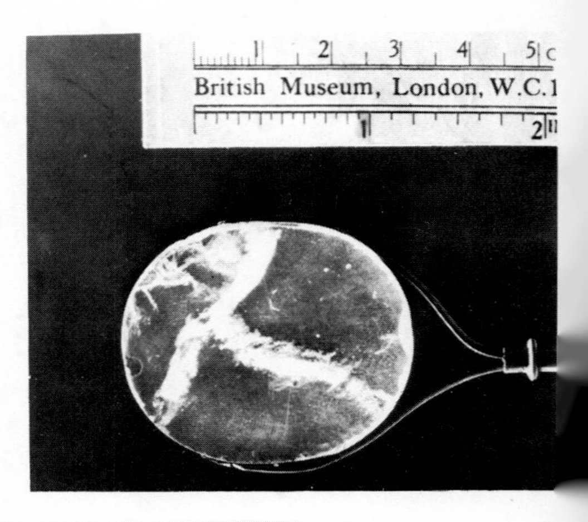

An Assyrian crystal lens from the seventh century B.C. To grind such a lens requires a highly sophisticated mathematical formula. Where did the Assyrians get such knowledge? Finally, these very ancient fragments, now in the Baghdad Museum, have been identified as the remains of an electric battery.

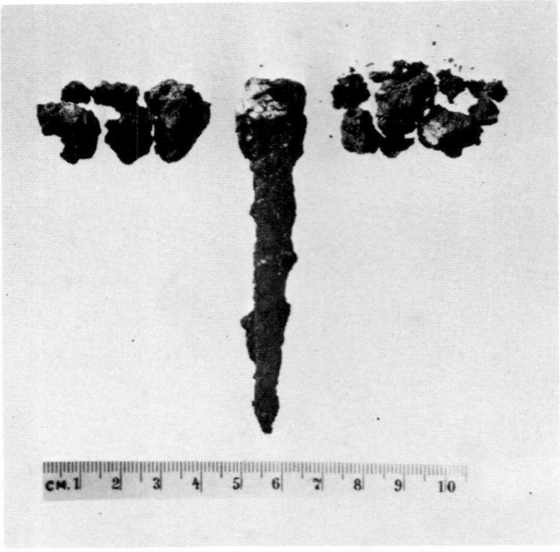

This temple at Copán in Honduras is constructed according to the Mayan calendar with a fixed number of steps completed every 52 years.

This ancient iron pillar does not rust. It cannot be accurately dated, but it is hundreds of years old.

This Babylonian tablet records past and future eclipses.

Rock drawings from all around the world. *Above left,* from Rhodesia. This reclining figure is clad in chain mail and wears curious headgear. It might be the burial of a king. It might just as likely be an astronaut receiving supplies. *Above right*: Drawing from South Africa showing a white figure dressed in a short-sleeved suit with breeches, garters, gloves, and slippers. A rather surprising example of imagination on the part of primitive natives who went about naked. *Below left*: Drawing found by a Russian expedition. *Below right*: Drawing from Val Camonica in Northern Italy shows yet again the extraordinary obsession primitive man had with figures in suits and unusual headgear.

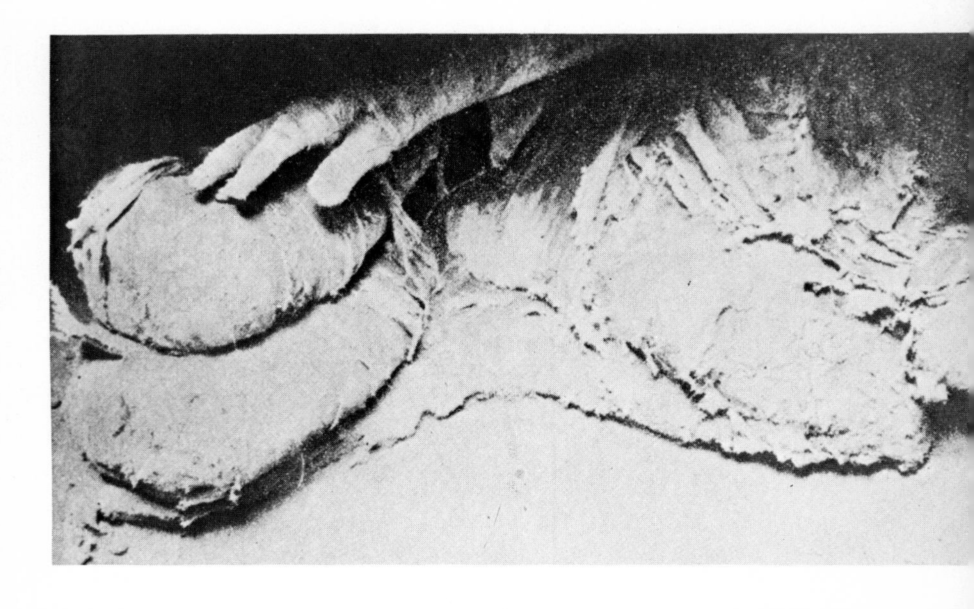

Above: A mummy from the Second Dynasty. *Below*: Part of a very finely woven piece of cloth. Where did the Egyptians get such complex techniques so early?

Above: With wood rollers and manpower it would have taken at least 600 years to handle the 2,500,000 stone blocks in the great pyramid of Cheops.

Below: As in Peru, we are faced with fantastically accurate jointing of huge blocks of stone.

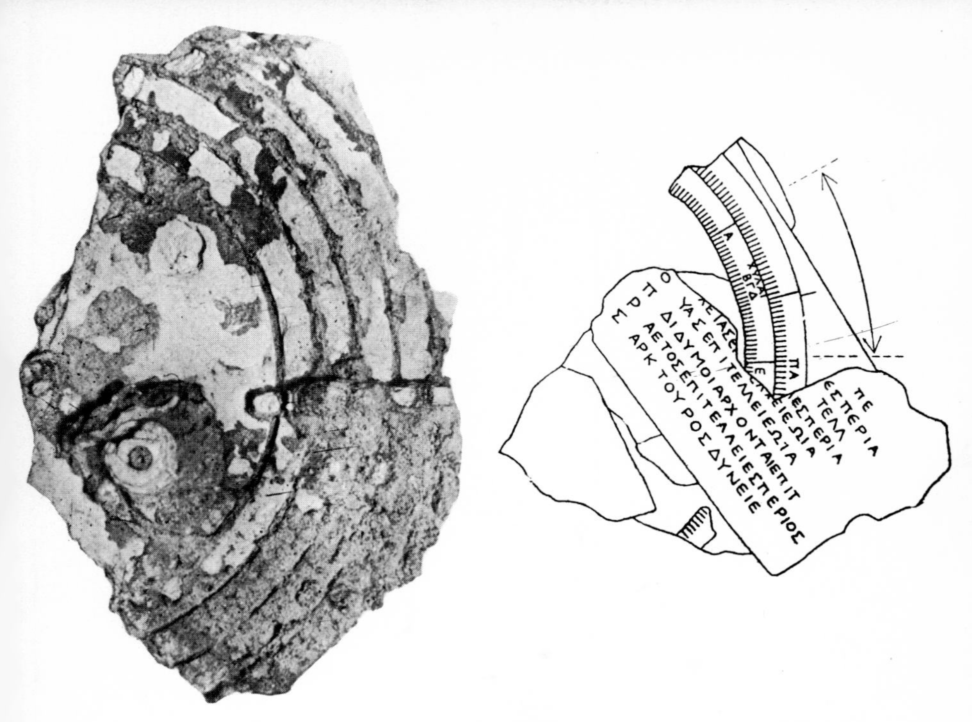

Above left: After centuries under the sea this object does not look very important. It was found by Greek divers off Antikythera in 1900. *Above right*: Long and patient cleaning revealed that it was a mass of interlocking cogs and was a planetarium. The machine gives the year of construction as 82 B.C. This drawing reconstructs part of the machine.

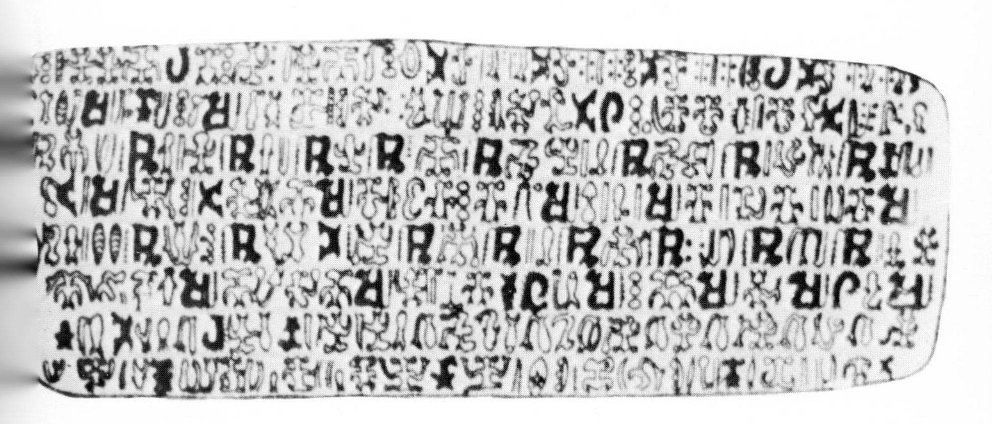

Easter Island is known as the "navel of the world" by the islanders. The huge statues on this tiny fragment of rock are extraordinary. Even more extraordinary is the fact that this totally isolated island had its own script, which is still undeciphered.

More ancient drawings. *Above,* from Navoy, and *below,* from Fergana in Uzbekistan.

An American astronaut. Perhaps the earliest space travelers also wore suits like this.

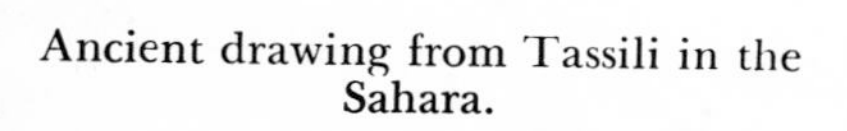

Ancient drawing from Tassili in the Sahara.

On July 21, 1969, a London bookmaker paid out £10,000. Only five years earlier he gave odds of a thousand to one that man would not land on the moon before 1970. Two weeks after the moon shot, the Americans released photographs taken 2,000 miles away from Mars. Space travel had moved out of the realm of science fiction.

stroyed by a great catastrophe exist in Death Valley, in the Nevada Desert. Even today traces of melted rocks and sand can be seen. The heat of a volcanic eruption would not have been enough to melt rocks—besides, the heat would have scorched the buildings first. Today only laser beams produce the required temperature. Strangely enough, not a blade of grass grows in this district.

Hadjar el Guble, the Stone of the South, in the Lebanon weighs more than 2,000,000 pounds. It is a dressed stone, but human hands could certainly not have moved it.

There are artificially produced markings, as yet unexplained, on extremely inaccessible rock faces in Australia, Peru, and Upper Italy.

Texts on gold plaques, which were found at Ur in Chaldea, tell of "gods" resembling men who came from the sky and presented the plaques to the priests.

In Australia, France, India, the Lebanon, South Africa, and Chile there are strange black "stones" which are rich in aluminum and beryllium. The most recent investigations showed that these stones must have been exposed to a heavy radioactive bombardment and high temperatures in the very remote past.

Sumerian cuneiform tablets show fixed stars with planets.

In Russia archaeologists discovered a relief of an airship, consisting of ten balls arranged in a row next to one another which stand in a right-angled frame supported on both sides by thick columns. Balls rest on the columns. Among other Russian finds there is a small bronze statue of a humanoid being in a bulky suit which is hermetically closed at the neck by a helmet. Shoes and gloves are equally tightly attached to the suit.

In the British Museum the visitor can read the past and future eclipses of the moon on a Babylonian tablet.

Engravings of cylindrical rocketlike machines, which are shown climbing skyward, were discovered in Kunming, the capital of the Chinese province of Yunnan. The engravings were on a pyramid which suddenly emerged from the floor of Lake Kunming during an earthquake.

How is anyone going to explain these and many other puzzles to us? When people try to dismiss the old traditions wholesale as false, erroneous, meaningless, and irrelevant, they are merely dodging the issue. It is equally unreasonable, when all is said and done, to lump all traditions together as inaccurate and then make use of them when it happens to suit one's purpose. I think that there is something cowardly about stopping one's eyes and ears to facts—or even hypotheses—simply because new conclusions might win men away from a pattern of thought that has become familiar.

Revelations take place hourly and daily all over the world. Our modern means of communication and transport spread discoveries all around the globe. Scholars of all disciplines should inquire into reports from the past with the same creative enthusiasm that they bring to contemporary research. The adventure of the discovery of our past has finished its first phase. Now the second fascinating adventure in human history begins with man's moving into the cosmos.

10

The Earth's Experience of Space

THE question whether space travel has any point has not yet been silenced in discussion. The partial or total meaninglessness of space research is supposed to be proved by the banal assertion that people should not poke around in the universe as long as there are still so many unsolved problems on earth.

As I am anxious not to enter into the realm of scientific argument unintelligible to the layman, I shall only give a few obvious and valid reasons for the absolute necessity of space research.

From time immemorial, curiosity and the thirst for knowledge have always been the driving force for continuing research on the part of man. The two questions, WHY did something happen? and HOW did it happen? have always been the spur to development and progress. We owe our present-day standard of living to the permanent unrest that they created. Comfortable modern means of transport have removed the hardships of journeys which our grandfathers still had to suffer; many of the rigors of manual labor have been noticeably alleviated by machines; new sources of energy, chemical preparations, refrigerators, various household appliances, etc., have completely liberated us from many activities that formerly could be done only by human hands. The creations of science have become not the curse but the blessing of mankind. Even its most terrifying offspring, the atom bomb, will turn out to be for the benefit of mankind.

Today science reaches many of its goals with seven-league boots. It took 112 years for photography to develop to the stage of a clear picture. The telephone was ready for use in 56 years, and only 35 years of scientific research were needed to develop radio to the point of perfect reception. But the perfecting of radar took only 15 years. The stages of epoch-making discoveries and developments are getting shorter and shorter; black-and-white television was on view after 12 years' research, and the construction of the first atom bomb took a mere 6 years. These are a few examples from 50 years of technical progress—magnificent and even a little frightening. Development will continue to reach its targets faster and faster. The next hundred years will realize the majority of mankind's eternal dreams.

The human spirit has made its way in the face of opposition and warnings. In the face of the archaic writing on the wall saying that water was the fishes' element and air the birds' element, man has conquered the regions which were not apparently intended for him. Man flies, against all the so-called laws of nature, and he lives under water for months in nuclear-powered submarines. Using his intelligence, he has made himself wings and gills which his creator had not intended for him.

When Charles Lindbergh began his legendary flight, his goal was Paris; obviously he was not really concerned with getting to Paris; he wanted to demonstrate that man could fly the Atlantic alone and unharmed. The first goal of space travel was the moon. But what this new scientific-cum-technical project really wants to prove is that man can also master space.

So why space travel?

In only a few centuries our globe will be hopelessly and irremediably overpopulated. Statistics already calculate a world population of 8.7 billion for the year 2050. Barely 200 years later it will be 50 billion, and then 335 men will have to live on one square kilometer. It doesn't bear thinking about! The tranquilizerlike theories of food from the sea or even cities on the floor of the sea will prove inefficient remedies against the population explosion sooner than their optimistic supporters would like to think. In the first six months of 1966 more than 10,000 people, who had tried in desperation to keep themselves alive by eating snails and plants, starved to death on the Indonesian island of Lombok. U Thant, Secretary General of the United Nations, estimates the number of children in danger of dying of hunger in India at 20,000,000, a figure which backs up Dr. Hermann Mohler of Zurich's claim that hunger is reaching for world domination.

It has been proved that world food production does not keep pace with the growth of population, in spite of the most modern technical aids and the large-scale use of chemical fertilizers. Thanks to chemistry, the present age also has birth control products at its disposal. But what use are they if the women in underdeveloped countries do not use them? For food production could draw level with the population increase only if it were possible to halve the birth rate in ten years, *i.e.*, by 1980. Unfortunately I cannot believe in this rational solution, because the "sound barrier" of prejudice, ostensibly due to ethical motives and religious laws, cannot be broken through as quickly as the calamity of overpopulation grows. Is it more human or even divine to let millions of people die of hunger year after year than to save the poor creatures from being born?

Yet even if birth control were to win through one fine day, even if cultivatable areas were enlarged and harvests multiplied by aids as yet unknown, even if fishing supplied much more food and fields of algae on the ocean bed provided nourishment, if all this and a lot more were to happen, it would all be only a postponement, a putting off of the evil day for about 100 years.

I am convinced that one day men will settle on Mars and cope with the climatic conditions just as the Eskimos would do if they were transplanted to Egypt. Planets, reached by gigantic spaceships, will be populated by our children's children; they will colonize new worlds, just as America and Australia were colonized in the comparatively recent past. That is why we must press on with space research.

We must bequeath our grandchildren a chance to survive. Every generation which neglects this duty is condemning the whole of mankind to death by starvation some time in the future.

It is no longer a question of abstract research which is of interest only to the scientist. And let me impress on anyone who does not feel that he is responsible for the future that the results of space research have already protected us from a third world war. Has not the threat of total annihilation prevented the great powers from settling opinions, challenges, and conflicts with a major war? It is not necessary now for a Russian soldier to set foot on American soil in order to transform the United States into a desert, and no American soldier need ever die in Russia, because an atom bomb attack makes a country uninhabitable and barren owing to radioactivity. It may sound absurd, but the first intercontinental missiles guaranteed us comparative peace.

The view is occasionally put forward that the billions invested in space research would be better spent on assisting development. This view is wrong; the industrial nations do not give aid to underdeveloped countries purely on charitable or political grounds; they also give it, understandably enough, to open up new markets for their own industries. The aid that the underdeveloped countries require is irrelevant from a long-term point of view.

Approximately 1.6 billion rats, each of which destroyed about 10 pounds of food a year, were living in India in 1966. Yet the state does not dare exterminate this plague, because the devout Indian protects rats. India also has a population of 80,000,000 cows, which give no milk, cannot be harnessed as draft animals, and cannot be slaughtered. In a backward country whose development is hindered by so many religious taboos and laws, it will take many generations to sweep away all the life-endangering rites, customs, and superstitions.

Here, too, the means of communication of the age of space travel—newspapers, radio, television—serve progress and enlightenment. The world has become smaller. We know and learn more about one another. But to arrive at the ultimate insight that national frontiers are a thing of the past, space travel was needed. The resulting increase in technology will spread the realization that the insignificance of peoples and continents in the dimensions of the universe can only be a stimulus and incentive to cooperative work on space research. In every epoch mankind has needed an inspiriting watchword that enabled it to rise beyond the obvious problems to the apparently unattainable reality.

A quite considerable factor which provides an important argument for space research in the industrial age is the appearance of new branches of industry, in which hundreds of thousands of people who lost their jobs through automation now earn their living. The space industry has already outstripped the automobile and steel industries as a pacesetter in the market. More than 4,000 new articles owe their existence to space research; they are virtually by-products of research for a higher goal. These by-products have become an accepted part of everyday life without anyone giving a thought about their origin. Electronic calculating machines, mini-transmitters and mini-receivers, transistors in radio and television sets, were discovered on the periphery of research, and so were the frying pans in which food does not stick. Precision instruments in all aircraft, fully automatic ground control systems and automatic pilots, and the rapidly developed computer are parts of the space research that has so many persecutors, parts of a development program, that also have an effect on the private lives of individuals. The things of which the layman has no idea are legion: new welding and lubricating processes in an absolute vacuum, photoelectric cells and new tiny sources of energy conquering infinite distances.

Out of the flood of taxes which is poured into space research, the returns on the vast investment flow back to the taxpayer in a steady stream. Nations that do not participate in space research in any way will be overwhelmed by the technical revolution. Names and concepts such as Telstar, Echo, Relay, Trios, Mariner, Ranger, and Syncom are signposts on the road of irresistible research.

Since terrestrial supplies of energy are not inexhaustible, the space travel program will also become vital one day, because we shall have to obtain fissionable matter from Mars or some other planet in order to be able to illuminate our cities and heat our houses. As atomic power stations provide the cheapest form of energy already today, industrial mass production will be fully dependent on these stations only when the earth no longer yields fissionable matter. Fresh consequences of research overwhelm us daily. The leisurely transmission of acquired knowledge from father to son is over forever. A technician who repairs a radio set that works by simply pressing a button must know all about the technology of transistors and complicated circuits that are often printed on sheets of plastic. It will not be long before he also has to deal with the tiny new components of microelectronics. What the apprentice is taught today, the journeyman will have to fill out with new knowledge. And even if the man who was master of his craft in the days of our grandfathers had knowledge to last his whole life, the master of the present and future will constantly have to keep on adding new skills to old. What was valid yesterday is obsolete tomorrow.

Even though it will take millions of years, our sun will burn out and die one day. It does not even need that terrible moment when a statesman loses his nerve and sets the atomic annihilation apparatus in motion to cause a catastrophe. An unascertainable and unpredictable cosmic event could bring about the earth's downfall. Man has never yet accepted the idea of such a possibility, and it may be for that reason that he devoutly sought the hope of an afterlife of the spirit and soul in one of the many thousand religions.

So I suggest that space research is not the product of his free choice but that he is following a strong inner compulsion when he examines the prospects of his future in the universe. Just as I proclaim the hypothesis that we received visits from space in the dim and distant past, I also assume that we are not the only intelligences in the cosmos—indeed I suspect that there are older, more advanced intelligences in the universe. If I now also assert that all the intelligences are carrying on space research on their own initiative, I am really moving into the world of science fiction for a moment, knowing full well that I am putting my head into a hornets' nest!

"Flying saucers" have been appearing on and off for at least twenty years; in the literature on the subject they are known as UFO's, unidentified flying objects. But before I deal with the exciting subject of the mysterious UFO's, I should just like to mention an important argument used when the justification for space travel is under discussion.

It is said that research into space travel is unprofitable; no country, however rich, can raise the enormous amounts of money needed without risking national bankruptcy. True, research per se has never been profitable; it is the products of research that make the investment profitable. It is unreasonable to expect profitableness and the amortization of research into space travel at its present stage. No balance has been struck to show the return from the 4,000 by-products of space research. To me there is absolutely no doubt that it will give a return such as has seldom been given by any other kind of research. When it reaches its goal, not only will it be profitable, but it will also bring the salvation of mankind from downfall in the literal sense of the word. Incidentally, a whole series of COMSAT satellites are already sound commercial propositions.

In November, 1967, the German magazine *Der Stern* said:

> The majority of medical life-saving machines come from America. They are the product of the systematic evaluation of the results of atomic research, space travel, and military technology. And they are the product of a novel collaboration between industrial giants and hospitals in America, which is leading medicine to new triumphs almost daily.
>
> Thus the Lockheed Company which makes Starfighters and the famous Mayo Clinic cooperated to develop a new system of nursing based on computer techniques. The designers of North American Aviation, following suggestions by the medical profession, are working on an "emphysema belt," which is intended to make it easier for patients with lung trouble to breathe. NASA space authorities have produced the idea for a diagnostic apparatus. The apparatus, actually conceived to measure the impact of micro-meteorites on spaceships, registers minute muscular spasms in certain nervous diseases.
>
> Another life-saving by-product of American computer technology was the "heart-beat machine." Today more than 2,000 Germans live with one of these apparatuses in their chests. It is a battery-driven mini-generator which is introduced under the skin. From it the doctors insert a connecting cable through the superior vena cava to the right auricle of the heart. The

heart is then stimulated to rhythmical movements by regular surges of current. It beats. When the battery of the "heart machine" is burned out after three years, it can be changed by a comparatively simple operation.

General Electric improved this little miracle of medical technology last year when it developed a two-speed model. If the wearer of this appliance wants to play tennis or run to catch a train, he simply moves a bar magnet up and down for a moment over the spot where his built-in generator is located. His heart promptly works at a higher speed.

Two more examples of by-products of space research. Who still has the nerve to say that it is useless?

Under the headline "Stimulus from Moon Rockets," the newspaper *Die Zeit* contained the following report in November, 1967:

> The designs of space vehicles developed for soft landings on the moon have an interim interest for automobile manufacturers, for the knowledge of how such designs behave under conditions which cause their destruction can be appreciably increased. Even though it will not be possible to make cars safe for the passengers against all kinds of collisions, the designs used with most success in space travel can help to diminish the risk when collisions occur. "Honey-comb" sheets, which are being used more and more in modern aircraft construction, guarantee high tensile strength with little weight. They have also been practically tested in automobile manufacture. The floor of the experimental gas-turbine-driven Rover car is made of "honey-combs."

Anyone who knows the present state of research and the impetuous way in which it develops can no longer tolerate sayings such as, "It will never be possible to travel from one star to another." The younger generation of our day will see this "impossibility" become reality. Gigantic spaceships with incredibly powerful motors will be built, as the Russians proved in 1967 when they succeeded in coupling two unmanned spacecraft in the stratosphere. One sector of space research is already working on a kind of protective screen, like an electric rainbow, which is attached in front of the actual capsule and is intended to prevent or deflect the impact of particles. A group of distinguished physicists is trying to detect what are known as tachyons, theoretical particles which are supposed to fly faster than light and whose lower speed limit is the speed of light. Scientists know that tachyons must exist; it is now "only" a matter of providing physical proof of their existence. Yet such proofs have actually been produced for neutrinos and antimatter! Finally I should like to ask the die-hard critics in the chorus of opponents of space travel: Do you really believe that several thousands of probably the cleverest men of our time would waste their impassioned work on a pure Utopia or a trivial goal?

So let me tackle UFO's boldly, ignoring the risk of not being taken seriously. If I am not taken seriously, I can at least console myself with the knowledge that I am in distinguished company.

UFO's have been sighted in America and over the Philippines, in West Germany and elsewhere. Let us assume that 98 percent of the people who claimed that they had seen UFO's actually saw ball lightning, weather balloons, strange cloud formations, new unknown types of aircraft, or even odd effects of light and shade in the sky at twilight. Undoubtedly, too, many people were the victims of mass hysteria. They claimed to have seen something that simply was not there. And of course there were also the publicity-seekers who wanted to make capital out of their alleged observations and produce banner headlines for the press in the silly season. If we reject all the crackpots, liars, hysterics, and sensation-mongers, there still remains a sizable group of sober observers, including people whose jobs make them familiar with celestial phenomena. A simple housewife may have made the same mistake as a farmer in the Wild West. But when, for example, a sighting of UFO's is made by an experienced airline pilot, it is hard to dismiss it as humbug. For an airline pilot is familiar with mirages, ball lightning, weather balloons, etc. The reactions of all his senses, including his first-class vision, are regularly tested; he is not allowed to drink alcohol for some hours before takeoff and during flights. And an airline pilot is hardly likely to talk nonsense, because he would lose his nice, well-paid job only too easily. Yet when not merely one airline pilot, but a whole group of pilots (including Air Force men), tell the same story, we are bound to listen to it.

I myself do not know what UFO's are; I do not say that they have been proved to be flying objects belonging to unknown intelligences, although there could be little objection to such a supposition. Unfortunately I have never seen a UFO with my own eyes during my worldwide travels, but I can reproduce here some credible, authenticated accounts:

On February 5, 1965, the U.S. Department of Defense announced that the Special Division for UFO's had been instructed to investigate the reports of two radar operators. On January 29, 1965, these two men had spotted two unidentified flying objects on their radar screen at the Naval Airfield in Maryland. These objects approached the airfield from the south at the enormous speed of 4,350 miles an hour. Thirty miles above the airfield the objects made a sharp turn and quickly disappeared out of radar range.

On May 3, 1964, various people at Canberra, Australia, including three meteorologists, observed a large shining flying object crossing the morning sky in a northeasterly direction. During an interrogation by delegates of NASA the eyewitnesses described how the "thing" had tumbled about in a strange way and how a smaller object had rushed at the large one. The small object had given off a red glow and then been obliterated, while the large "thing" had disappeared from view in a northwesterly direction. One of the meteorologists said resignedly, "I've always ridiculed these UFO stories. What the hell am I going to say now?"

On November 23, 1953, an unidentified flying object was picked up on the radar screen of the Kinross Air Base in Michigan. Flight Lieutenant R. Wilson, who happened to be on a training flight in an F-86 jet aircraft, was given permission to chase the "thing." The radar crew watched Wilson pursuing the unidentified object for 160 miles. Suddenly both flying bodies merged with one another on the radar screen. Radio calls to Wilson were unanswered. During the next few days, the region in which the inexplicable event took place was combed for wreckage by search troops, and nearby Lake Superior was examined for traces of oil. They found nothing. There was absolutely no trace of Flight Lieutenant Wilson and his machine!

On September 13, 1965, shortly before one in the morning, Police Sergeant Eugene Bertrand came across a distracted woman at the wheel of her car in a bypass at Exeter, New Hampshire. The lady refused to drive on and claimed that a gigantic gleaming-red flying object had pursued her for ten miles to Route 101 and then disappeared into the forest.

The policeman, an elderly, level-headed man, thought the lady was a bit crazy, until he heard the same report from another patrol over his car radio. Speaking from headquarters, his colleague Gene Toland ordered him to return there at once. There a young man told him the same story as the lady; he too had sought refuge in the ditch from a glowing red object.

Rather unwillingly the men went on a car patrol, convinced that the whole silly story would have a rational explanation. They searched the district for two hours, then they set off on the return journey. They passed a field in which stood six horses that suddenly stampeded madly out of it. Almost simultaneously the region was bathed in glowing red light. "There. Look there!" shouted a young policeman. Indeed, a fiery red object, which moved slowly and silently toward the observers, was floating above the trees. Bertrand excitedly informed his colleague Toland over the telephone that he had just seen the damned thing with his own eyes. Now the farm near the road and the neighboring hill were also bathed in glowing red light. A second police car screeched to a halt next to the men.

"God damn it!" stuttered Dave. "I heard you and Toland yelling to each other over the radio. I thought you'd gone crazy. But just look at that!"

Fifty-eight qualified eyewitnesses came forward during the investigation of the mysterious incident that was subsequently carried out. They included meteorologists and members of the Coast Guard—in other words, men who as reliable observers were scarcely likely not to be able to tell a weather balloon from a helicopter, or a falling satellite from the navigation lights of an aircraft. The report contained factual statements but did not give any explanation of the unidentified flying object.

On May 5, 1967, the mayor of Marliens in the Côte-d'Or, Monsieur Malliotte, discovered a strange hole in a field of clover 680 yards from the road. He found traces of a circle with a diameter of 15½ feet and a depth of 1 foot. Deep furrows 4 inches deep ran out in all directions from this circle. They gave the impression that a heavy metal grating had been pressed into the ground. At the end of the furrows were holes 1 foot 2 inches deep, which might have been impressed in the soil by "feet" at the end of the metal grating. An exceptionally curious feature was the violet-white dust which was deposited in the furrows and holes. I have inspected this place near Marliens personally. Ghosts could not have left those traces!

What are we to make of this account? It is depressing what

many people—and sometimes whole occult societies—make out of their ostensible observations. They only blur our view of reality and deter serious scholars from dealing with verified phenomena because they are afraid of exposing themselves to ridicule.

On November 6, 1967, during a transmission by German television on the subject "Invasion from the Cosmos?" the captain of a Lufthansa aircraft told of an incident of which he and four members of the crew were eyewitnesses. On February 15, 1967, about ten to fifteen minutes before landing in San Francisco, they saw close to their own machine a flying object with a diameter of about 33 feet that shone dazzlingly and flew alongside them for some time. They sent their observations to the University of Colorado, which for want of a better explanation surmised that the flying object was part of a previously launched rocket falling to the ground. The pilot explained that with more than a million miles of flying experience he was as unable as his colleagues to believe that a falling piece of metal could stay in the air for a quarter of an hour, have such dimensions, and fly alongside an aircraft; he believed this explanation even less since this unidentified flying object had been observable from the ground for nearly three-quarters of an hour. The German pilot certainly did not give the impression of being a visionary.

Two reports from *Die Suddeutsche Zeitung,* Munich, November 21 and 23, 1967:

> *Belgrade* (*From our own correspondent*): Unidentified flying objects (UFO's) have been sighted over various districts of southeast Europe during the last few days. At the weekend an amateur astronomer photographed three of these gleaming celestial objects at Agram. But while the experts were still giving their opinions of this photograph that was splashed across several columns of the Yugoslavian papers, more UFO's have already been reported from the mountainous region of

Montenegro, where they were even supposed to have caused several forest fires. These accounts come mainly from the village of Ivangrad where the inhabitants swear black and blue that they have observed strange brightly illuminated heavenly bodies every evening during the last few days. The authorities confirm that several forest fires have occurred in this district but so far cannot explain what started them.

Sofia (*UPI*): A UFO has appeared over the Bulgarian capital of Sofia. According to the report of the Bulgarian News Agency BTA, the UFO could be recognized with the naked eye. BTA says that the flying body was "bigger than the sun's disc and later took the shape of a trapeze." The flying body is supposed to have emitted powerful rays. It was also observed by telescope in Sofia. A scientific collaborator of the Bulgarian Institute for Hydrology and Meteorology said that the flying body apparently moved under its own power. It was flying about 18 miles above the earth.

People block the road to serious research by boundless stupidity. There are "contact men" who claim to be in communication with extraterrestrial beings; there are groups who develop fanciful religious ideas from hitherto unexplained phenomena or build cranky philosophies of life from them or even claim to have received orders for the salvation of mankind from UFO crews. Among the religious fanatics, the Egyptian "UFO angel" naturally comes from Mohammed, the Asiatic one from Buddha, and the Christian one directly from Jesus.

At the 7th International World Congress of UFO Investigators, in the autumn of 1967, Professor Hermann Oberth, the man known as "the father of space travel" and the teacher of Wernher von Braun, said that UFO's were still "an extra-scientific problem"; but, said Oberth, UFO's were probably "spaceships from unknown worlds," and to use his own words: "Obviously the beings who man and fly them are far ahead of us culturally, and if we go about things properly we can learn a lot from them." Oberth, who accurately predicted rocket development on earth, suspects that the prerequisites for abiogenesis exist on other planets in the solar system. Oberth, a research scientist himself, demands that serious scientists, too, should tackle problems that may seem fantastic at first. "Scholars behave like stuffed geese who refuse to digest anything else. They simply reject new ideas as nonsense."

On November 17, 1967, under the headline "Second Thoughts," *Die Zeit* said:

> For years the Russians have ridiculed Western hysteria about flying saucers. Not long ago *Pravda* contained an official denial that such peculiar celestial vehicles existed. Now Air Force General Anatolyi Stolyakov has been appointed director of a committee which is to examine all reports of UFO's. In this connection the London *Times* writes: "Whether UFO's are the product of collective hallucinations, whether they originate from Venusian visitors or are to be understood as a divine revelation—there must be an explanation for them, otherwise the Russians would never have set up a Committee of Inquiry."

The most spectacular and puzzling incident connected with the phenomenon of "matter from the universe" took place at 7:17 on the morning of July 30, 1908, in the Siberian Taiga. A fireball shot across the sky and was lost in the steppe. Travelers on the Trans-Siberian Railway observed a glowering mass which moved from south to north. A thunderbolt shook the train, explosions followed, and most of the seismographic stations in the world registered an appreciable earth tremor. At Irkutsk, 550 miles from the epicenter, the needle of the seismograph went on quivering for nearly an hour. The noise could be heard over a radius of 621 miles. Whole herds of reindeer were destroyed. Nomads were whirled up into the air with their tents.

Not until 1921 did Professor Kulik begin to collect eyewitness accounts. Finally he also succeeded in collecting the money for a scientific expedition to this sparsely populated region of the Taiga.

When the expedition members reached the stony Tunguska in 1927, they were convinced that they would find the crater made by a gigantic meteorite. Their conviction turned out to be quite wrong. They saw the first trees without tops as much as 37 miles from the center of the explosion. The nearer they came to the critical point, the more barren the district became. Trees stood there like shaved telegraph poles; in the vicinity of the center even the strongest trees had been snapped off outward. Lastly they found traces of a tremendous conflagration. Pushing on farther north, the expedition became convinced that a vast explosion must have taken place. When they came across holes of all sizes in swampy ground they suspected the impact of meteorites; they dug and drilled in the marshy ground without finding a single remnant, a piece of iron, a bit of nickel, or a lump of stone. Two years later the search was continued with bigger drills and improved technical resources. They drilled to a depth of 118 feet without finding a single trace of any kind of meteoric material.

In 1961 and 1963 two more expeditions were sent to the Tunguska by the Soviet Academy of Sciences. The 1963 expedition was under the leadership of the geophysician Solotov. This group of scientists, now equipped with the most modern technical appliances, came to the conclusion that the explosion in the Siberian Tunguska must have been a nuclear one.

The type of an explosion can be determined when several physical orders of magnitude that caused it are known. One of these orders of magnitude in the Tunguska explosion was known in the vast amount of radiant energy emitted. In the Taiga the expedition found trees 11 miles from the center of the explosion that had been exposed to radiation and set on fire by it at the moment of explosion. But a growing tree can catch fire only if the amount of radiant energy per square centimeter reaches 70 to 100 calories. Yet the flash of the explosion was so bright that it continued to cast secondary shadows at a distance of 124 miles from the epicenter!

From these measurements the scientists worked out that the radiant energy of the explosion must have been around 2.8×10^{23} ergs. (The erg in science is the so-called "measurement of work." A beetle weighing one gram performs 1 erg's worth of work when it climbs a wall 1 centimeter high.)

The expedition found branches and twigs on the tops of trees that had been carbonized, up to a range of eleven miles. From this they concluded that sudden heating had taken place. This was the result of an explosion, not a forest fire! These carbonizations were found only where there had been no shadows to interrupt the diffusion of the flash. Clearly and unquestionably it must have been a case of radiation. The sum of all these effects makes the force of 10^{23} ergs necessary for the gigantic devastation. This immense energy corresponds to the destructive power of an atom bomb weighing 10 megatons or

100,000,000,000,000,000,000,000 ergs!

The investigations confirmed a nuclear explosion and relegated to the realm of fable explanations such as the impact of a comet or the fall of a great meteorite.

What explanations are offered for this nuclear explosion in the year 1908?

In March, 1964, an article in the reputable Leningrad paper *Svesda* put forward the theory that intelligent beings living on a planet in the constellation Cygnus had tried to make contact with the earth. The authors, Genrich Altov and Valentina Shuraleva, said that the impact in the Siberian Taiga was a response to the colossal explosionlike eruption of the volcano of Krakatoa in the Indian Ocean, which sent a large concentration of radio waves into the universe when it erupted in 1883. The distant stellar beings had erroneously taken the radio waves for a signal from space; so they had directed a laser beam, which was much too strong, at the earth, and when the beam hit the earth's atmosphere high above Siberia, it had turned into matter. I must admit that I do not accept this explanation because it seems too farfetched.

I am equally unable to accept the theory that seeks to explain the incident by the impact of antimatter. Even though I believe that there is antimatter in the depths of the cosmos, there cannot be any left in the Tunguska, because the collision of matter and antimatter results in their mutual dissolution. Moreover, the possibility of a piece of antimatter reaching the earth without a collision with matter on its long journey is very remote.

I prefer to adhere to the opinion of those who suspect that the nuclear explosion was caused by an unknown spaceship's energy pile bursting. Fantastic? Of course, but does that mean that it must be impossible?

There are shelves and shelves of literature about the Tunguska meteorites. One further fact I want to emphasize: radioactivity around the center of the explosion in the Taiga is twice as high—even today—as elsewhere. Careful investigation of trees and their annual rings confirm an appreciable increase in radioactivity since 1908.

Until a single, exact, indubitable scientific proof of the phenomenon—and many others—is produced, no one has the right to discard an explanation within the bounds of credibility without giving his reasons.

Our knowledge of the planets in our solar system is rather comprehensive; Mars is the only planet on which "life" in our sense of the word might exist and then only in limited quantities. Man has set the theoretical boundary to the possibility of life in this sense; this boundary is called the ecosphere. In our solar system only Venus, the Earth, and Mars lie within the limits of the ecosphere. Nevertheless, we should remember that the determination of the ecosphere is based on our conception of life and that unknown life is by no means necessarily bound to our premises for life. Until 1962 Venus was considered to be a possible home for life. Then Mariner II got within about 21,000 miles of Venus. According to the information it transmitted, Venus can now be ruled out as a supporter of life.

It emerged from Mariner II's reports that the average surface temperature on both light and dark sides was 420° C. Such a temperature means that there could be no water, only lakes of molten metal on the surface. The popular idea of Venus as the twin sister of the earth is over and done with, even though the carbureted hydrogen present could be a culture medium for all kinds of bacteria.

It is not long since scientists claimed that life on Mars is inconceivable. For some time now that has become "is *scarcely* conceivable." For after the successful reconnaissance mission by Mariner IV we must concede, even if reluctantly, that the possibility of life on Mars is not unlikely. It is also within the bounds of possibility that our neighbor Mars had its own civilization untold millennia ago. In any case the Martian moon Phobos deserves special attention.

Mars has two moons: Phobos and Deimos (in Greek, Fear and Terror). They were known long before the American astronomer Asaph Hall discovered them in 1877. As early as 1610 Johannes Kepler suspected that Mars was accompanied by two satellites. Although the Capucine monk Schyrl may have claimed to have seen the Martian moons a few years earlier, he must have been mistaken, for the tiny Martian moons could not possibly have been seen with the optical instruments of his day. A fascinating description of them is given by Jonathan Swift in *A Voyage to Laputa and Japan*, which forms Part III of *Gulliver's Travels*. Not only does he describe the two Martian moons, but he also gives their size and orbits. This quotation comes from Chapter 3:

> [The Laputan astronomers] spend the greatest part of their lives in observing the celestial bodies, which they do by the assistance of glasses far excelling ours in goodness. For although their largest telescopes do not exceed three feet, they magnify much more than those of a hundred yards among us, and at the same time show the stars with greater clearness. This advantage hath enabled them to extend their discoveries much further than our astronomers in Europe for they have made a catalogue of ten thousand fixed stars, whereas the largest of ours do not contain above one third part of that number. They have likewise discovered two lesser stars, or satellites, which revolve about Mars, whereof the innermost is distant from the centre of the primary planet exactly three of the diameters, and the outermost five; the former revolves in the space of ten hours, and the latter in twenty one and an half; so that the squares of their periodical times are very near in the same proportion with the cubes of their distance from the centre of Mars, which evidently shows them to be governed by the same law of gravitation, that influences the other heavenly bodies.

How could Swift describe the Martian satellites when they were not discovered until 150 years later? Undoubtedly the Martian satellites were suspected by some astronomers before Swift, but suspicions are not nearly enough for such precise data. We do not know where Swift got his knowledge.

Actually these satellites are the smallest and strangest moons in our solar system. They rotate in almost circular orbits above the equator. If they reflect the same amount of light as our moon, then Phobos must have a diameter of 10 miles and Deimos one of only 5 miles. But if they are artificial moons and so reflect still more light, they would actually be even smaller. They are the only known moons in our solar system that move around their mother planet faster than she herself rotates. In relation to the rotation of Mars, Phobos completes two orbits in one Martian day, whereas Deimos moves only a little faster around Mars than the planet itself rotates.

In 1862, when the earth was in a very favorable position in relation to Mars, people sought in vain for the Martian satellites—they were not discovered until fifteen years later! The theory of planetoids came up because several astronomers suspected that the Martian moons were fragments from space which Mars had attracted. But the theory of planetoids is untenable, for both the Martian moons revolve in almost the same planes above the equator. *One* fragment from space might do that by chance, but not two. Finally, measurable facts produced the modern satellite theory.

Russian scientist I. S. Shklovskii and renowned American astonomer Carl Sagan, in their book *Intelligent Life in the Universe,* published in 1966, accept that the moon Phobos is an artificial satellite. As the result of a series of measurements, Sagan came to the conclusion that Phobos must be hollow and a hollow moon cannot be natural.

In fact, the peculiarities of Phobos' orbit bear no relation to its apparent mass, whereas such orbits are typical in the case of hollow bodies. Shklovskii, director of the Department of Radio-Astronomy in the Moscow Sternberg Astrological Institute, made the same statement after he had observed that a peculiar unnatural acceleration could be confirmed in the movement of Phobos. This acceleration is identical with the phenomenon which has been established in the case of our own artificial satellites.

Today people take these fantastic theories of Sagan and Shklovskii very seriously. Further Martian probes are planned, also intended to take the bearings of the Martian moons. In the years ahead, the Russians intend to observe the movements of the Martian moons from several observatories.

If the view supported by reputable scientists East and West that Mars once had an advanced civilization is correct, the question arises: Why does it no longer exist today? Did the intelligences on Mars have to seek a new environment? Did their home planet, which was losing more and more oxygen, force them to look for new territories to settle? Was a cosmic catastrophe responsible for the downfall of the civilization? Lastly, were some of the inhabitants of Mars able to escape to a neighboring planet?

In his book *Worlds in Collision,* published in 1950 and still much discussed in scientific circles, Immanuel Velikovsky declared that a giant comet had crashed into Mars and that Venus had been formed as a result of this collision. His theory can be proved if Venus has a high surface temperature, clouds containing carbureted hydrogen, and an anomalous rotation. Evaluation of the data provided by Mariner II confirms Velikovsky's theory. Venus is the only planet which rotates "backward," *i.e.,* the only planet that does not follow the rules of our solar system as do Mercury, the Earth, Mars, Jupiter, Saturn, Uranus, and Neptune.

But if a cosmically caused catastrophe is a possible reason for the destruction of a civilization on the planet Mars, that would also provide material for my theory that the earth may have received visits from space in the very remote past. The thesis that a group of Martian giants perhaps escaped to earth to found the new culture of *homo sapiens* by breeding with the semi-intelligent beings living there then becomes a speculative possibility. Since the gravity of Mars is not as strong as that of the earth, it can be assumed that the build of Martian men was heavier and bigger than that of the earth men. If there is anything in this argument, we could have the giants who came from the stars, who could move enormous blocks of stone, who instructed men in arts still unknown on earth, and who finally died out.

Never have we known so little about so much as today. I am certain that the theme Man and Unknown Intelligences will remain on the agenda of research until every puzzle that can be solved has found an answer.

11

The Search for Direct Communication

AT 4 o'clock one April morning in 1960, an experiment began in a lonely valley in West Virginia. The big 85-foot radio-telescope at Green Bank was trained on the star Tau Ceti, 11.8 light-years away. Young American astronomer Dr. Frank Drake, who enjoys considerable fame as a scientist, acted as leader of this project. He wanted to tune in to the radio transmissions of other civilizations in order to pick up signals from unknown intelligences in outer space. The first series of experiments lasted 150 hours. They passed into history as Project Ozma (after the princess of the mythical land of Oz), although it was a failure. The experiment was broken off, not because some of the participating scientists expressed the view that there were no radio transmissions in space, but rather because it was realized that at the time there was no apparatus sensitive enough to reach the desired goal. Ozma will not be the only experiment of its kind. Instruments have been erected on the moon, as of July, 1969, and more are to come in subsequent visits by astronauts. They will be able to scan the immeasurable spaces between the stars for radio signals, free from terrestrial interference.

However, it must be asked whether the search for radio signals really helps our space research and whether it might not be more practical for us to do the sending of radio signals into space. Of course, we cannot expect unknown intelligences to understand Russian or Spanish or English and to be sitting there waiting to be contacted.

There remain three possibilities by which we can make ourselves known: mathematical symbols, laser beams, and pictures. The first of these seems most likely to succeed. In order to send such symbols we shall have to discover and fix intergalactic wavelengths that stand a good chance of being received throughout the cosmos. 1420 megaherz would provide such a frequency, for that is the radiation frequency of the neutral hydrogen that results from the collision of hydrogen atoms. Since hydrogen is an element, this radiation frequency could be known throughout the universe. Besides, 1420 megaherz lies outside the overcrowded scale of terrestrial wavelengths. The possibility of errors and interference factors would be reduced to a minimum. In this way radio impulses could be sent into space and if unknown intelligences exist they would recognize them.

In this connection a news item from *Die Zeit,* December 22, 1967, is most interesting in the light of the actual moon landing in July, 1969. Under the headline "The Moon Will Be Bombarded by Flashes," we read:

> The distance of the moon from the earth is known to the nearest few hundred yards, but astronomers refuse to be satisfied with that. So astronauts on one of the first flights to the earth's satellite will take mirrors with them and set them up there. These mirrors—like the corner of a room—will consist of three reflecting planes standing at right angles to each other and will have the property of returning any light that strikes them back to the source of the light.

This mirror system will be bombarded from the earth by a laser emitting flashes of light each lasting for a hundred millionth of a second. The laser will be used in conjunction with a telescope with an aperture of 1.50 meters. The light reflected from the moon will be picked up by this telescope and led to a photo-copier.

The distance of the moon can then be determined to one and a half meters from the known speed of light and the time taken by a laser beam for the journey there and back.

The same kind of thing is also conceivable in reverse. Radio waves have been traversing the universe for a very long time. If my hypothesis is correct, isn't it credible that unknown intelligences are also announcing themselves to us? For example, the radiation energy of CTA 102 suddenly increased in the autumn of 1964; Russian astronomers informed the world that they had possibly received signals from an extraterrestrial supercivilization. (Radio star CTA 102 was listed under catalog number 102 by radio astronomers of the California Institute of Technology—hence its name.)

The astronomer Sholomitski said in the lecture room of the Sternberg Astrological Institute in Moscow on April 13, 1965: "At the end of September and the beginning of October, 1964, the radiation energy from CTA 102 was much stronger, but only for a short time, then it diminished again. We registered this and waited. Toward the end of the year the intensity of the source suddenly increased again; it reached a second peak exactly 100 days after the first record was taken." His chief, Professor I. S. Shklovskii, added that such fluctuations in radiation were very unusual.

Meanwhile Dutch astrophysicist Maarten Schmidt has found out by exact measurements that CTA 102 must be about 10 billion light-years from the earth. That means that if the radio beams originated from intelligent beings, they must have been radiated 10 billion years ago. But according to the calculations of present-day research, our planet simply did not exist at that time. This realization could mean a kind of *coup de grâce* to the search for other living beings in the universe.

But if the search for life in the universe had no chance of success, astrophysicists in America and Russia, at Jodrell Bank, England, and at Stockert near Bonn in Germany would not be concentrating their research on what are known as radio stars and quasars with enormous directional antennae. The fixed stars Epsilon Eridiani and Tau Ceti are respectively 10.2 and 11.8 light-years away from us. So radio waves aimed at these "neighbors" would be about 11 light-years under way, and an answer from them could reasonably reach us in 22 years. Radio communications with more distant stars would take correspondingly longer; civilizations situated at distances reckoned in millons of light-years are quite unsuitable to contact by means of radio waves. But are radio waves our only technical means of making such attempts?

For example, we could also make ourselves optically noticeable. A powerful laser beam directed at Mars or Jupiter could not remain unnoticed, provided that intelligent living beings are in existence there. ("Laser" is the abbreviation for "light amplification by the stimulated emission of radiation.") Another, somewhat fantastic-sounding possibility would be to cultivate vast areas of soil so that tremendous color contrasts appeared which at the same time represented a geometrical or mathematical symbol of universal validity. One audacious but perfectly realizable idea: a gigantic equilateral triangle would have its 600-mile-long sides sown with potatoes; in this enormous triangle a circle would be sown with wheat. In this way a vast yellow circle, surrounded by a green equilateral triangle, would appear every summer. Incidentally, a most useful and productive experiment! But if there are unknown intelligences that seek us as we seek them, the coloring of circle and triangle would be a hint to them that these shapes were no freaks of nature. Someone has also advocated erecting a chain of lighthouses which radiate their lights vertically. The resultant sea of light should be arranged to have the shape of a model of an atom.

There are all kinds of suggestions based on the premise that someone somewhere is watching our planet. Are we tackling the problem the wrong way by limiting ourselves to the kind of means suggested above?

However skeptical or antipathetic we may be to everything occult, we cannot avoid looking into some as yet inexplicable physical phenomena, for example the thought transference between intelligent brains that is proved on a broad scientific basis but not yet explained.

In the parapsychological departments of many important universities, previously unexplained phenomena such as clairvoyance, visions, and thought transference are being investigated with accurate scientific methods. In the process all ghost and bogey stories from dubious occult sources or inspired by religious mania are separated and rejected. In this field of research, which was absolutely taboo until quite recently, we have made important advances.

In August, 1959, the *Nautilus* experiment came to an end. It not only demonstrated the possibility of thought transference but also showed that mental communication between human brains can be stronger than radio waves. This was the experiment:

Thousands of miles away from the "thought transmitter," the submarine *Nautilus* dived several hundred feet below the surface. All radio communications were interrupted, for even today radio waves do not penetrate to any appreciable depth. On the other hand, mental communication between Mr. X and Mr. Y did function.

After such scientific tests one asks oneself what else the human brain is capable of. Can it make mental communications faster than light? The Cayce affair, which has passed into the annals of scientific literature, stimulates such suppositions.

Edgar Cayce, a simple farmer's son from Kentucky, had no idea of the fantastic capabilities that were hidden in his brain. Although he died on January 5, 1945, doctors and psychologists are still busy evaluating his actions. The strict American Medical Association gave Edgar Cayce permission to hold consultations, although he was not a doctor.

Edgar Cayce fell ill in his early youth; he was wracked by cramps; high fever was consuming his body; he fell into a coma. While the doctors were trying in vain to bring the boy back to consciousness, Edgar suddenly began to speak loudly and clearly. He explained why he was ill, named some remedies which he needed, and told them to prepare a paste from certain ingredients and smear his spine with it. Doctors and relatives were astounded because they had no idea where the boy had got this knowledge and the technical language, which was quite unknown to him. Edgar got progressively and visibly better after treatment with the medications he had named.

The incident was the talk of the state. Since Edgar had spoken in a coma, many people suggested that he should be hypnotized in order to "entice" suggestions for cures from him. Edgar would not have this at any price. Not until a friend of his fell ill did he dictate a precise prescription using Latin words which he had never heard or even seen before. A week later his friend was better again.

If the first case was soon forgotten as a minor sensation that was not to be taken seriously scientifically, the second incident caused the AMA to set up a commission which was to make reports if anything of the kind happened again and to put down in writing every single detail. In a sleeping state Cayce had knowledge and abilities which would normally be the result only of much consultation.

Once Cayce "prescribed" for a very wealthy patient a medicine that could not be procured anywhere. This man put several advertisements in widely circulating newspapers, including international ones. A young doctor wrote from Paris, saying that his father had made the medicine years ago but that production had long been discontinued. The composition of this medicine was identical with the detailed ingredients supplied by Edgar Cayce.

Later Cayce "prescribed" a medicine and also named the address of a laboratory in a town a long way away. A telephone call showed that the preparation was just being developed. A formula had been worked out, they were looking for a name, but it was not yet on sale to chemists.

The professional doctors on the commission were no believers in telepathy; they investigated soberly and objectively, verified what they observed, and knew that Cayce had never had a medical book in his hands in his life. Besieged on all sides and from all over the world, Cayce gave two consultations a day, always in the presence of doctors and always without accepting fees. His diagnoses and therapeutical prescriptions were accurate, but when he came out of his trance, he could not remember what he had said. When doctors on the commission asked him how he arrived at his diagnoses, Cayce supposed that he could put himself in contact with any brain required and gather the information he needed for his diagnoses from it. But as the patient's brain knew exactly what his body lacked, it was all very simple. He asked the brain of the sick person and then he sought out the brain in the world which could tell him what should be done. He himself, declared Cayce, was only a part of all brains.

An astonishing idea, which—transferred to the realm of technology—would look something like this. In New York a monster computer would be fed with all the known data on physics. Whenever and from wherever the computer was interrogated, it would give its answer in fractions of a second. Another computer might be in Zurich with the whole of medical knowledge stored inside it. One in Moscow would be stuffed with all the facts about biology. Another in Cairo would have no gaps in its astronomical knowledge. In short all the knowledge in the world, arranged by branches, would be stored in various centers. Connected by radio, the computer in Cairo, if asked for medical information, would pass on the questions to the computer in Zurich in a hundredth of a second. Edgar Cayce's brain must have functioned in much the same way as this perfectly credible and already technically feasible computer linkup.

I now put forward the bold speculation: What if all (or even only a few highly trained) human brains have unknown forms of energy at their disposal and possess the ability to make contact with all living beings? We know frighteningly little about the functions and potentialities of the human brain; but it is known that only one-tenth of the cortex functions in the brain of a healthy man. What are the remaining nine-tenths doing? The fact that men have recovered from incurable diseases by willpower and nothing else is well known and scientifically documented. Perhaps a "gear" unknown to us has been engaged, setting an additional tenth or two-tenths of the cortex working. If we assume the fantastic idea that the strongest forms of energy operate in the brain, then a strong mental impulse would be noticeable everywhere simultaneously. If science succeeds in making such a "wild" idea demonstrable, it could mean that all intelligences in the universe belong to the same unknown structure.

Let me give an example. If a strong electrical impulse is released at any point in a tank full of millions of bacteria, it is felt everywhere and by every species of bacteria. The surge of current is perceived everywhere at the same moment. I quite realize that this comparison is imperfect, for electricity is a known form of energy and dependent on the speed of light. I am concerned with a form of energy that is available and effective everywhere simultaneously. I imagine an as yet unidentified form of energy which will one day make the incomprehensible comprehensible.

In order to give a semblance of probability to the extraordinary idea, I shall quote the report of an experiment carried out May 29 and 30, 1965. In its scope and nature it must be unique. On these two days 1,008 people concentrated at the same time, indeed at the same second, on pictures, sentences, and groups of symbols, which were "radiated" into the universe by them with concentrated power. The fact of this mass experiment is not the only astonishing thing—the results are strange, too. None of the participants knew any of the others; they lived hundreds of miles apart. Yet 2.7 percent of the participants answered on forms that they had seen a picture, namely the model of an atom. Since collusion on the part of the "guinea pigs" was impossible, it is surprising that as many as 2.7 percent should have seen the same "mental picture." Telepathy? Hocus-pocus? Chance? Admittedly, the whole thing is a science-fiction subject, but the experiment, organized by scientists, did take place. It is quite obvious that we don't know everything yet. The result of a experiment by a group of physicists at Princeton University is equally inexplicable. While investigating the disintegration of electrically neutral K mesons, they reached a result that was theoretically impossible because it contradicted a long-established principle of nuclear physics.

One more extraordinary example. Einstein's theory of relativity says that mass and energy are only different forms of one and the same phenomenon ($E = mc^2$). Put simply, mass can literally be produced from the void. If a strong beam of energy is shot past a heavy atomic nucleus, the beam of energy disappears into the strong electrical field of energy of the atomic nucleus and an electron and a positron appear in its place. Energy in the form of a beam has changed into the mass of two electrons. To the mind that has not been trained scientifically the process seems crazy, and yet it takes place exactly like that. There is nothing to be ashamed of if you cannot follow Einstein; one scientist called him the great solitary because he could discuss his theory with only a dozen or so of his contemporaries.

After this excursion into the still unexplored fields of thought transference and the functions of the human brain, let us turn back to our theme again.

It is no longer a secret that in November, 1961, in the National Radio Astronomy Observatory at Green Bank, West Virginia, eleven authorities met at a secret conference. Here, too, the theme of the conference was the question of the existence of extraterrestrial intelligences. The scientists, among them Giuseppe Cocconi, Su Shu Huang, Philip Morrison, Frank Drake, Otto Struve, and Carl Sagan, as well as the Nobel Prize winner Melvin Calvin, collaborated at the end of the conference on what is known as the Green Bank Formula. According to this formula there are at any moment in our galaxy alone 50,000,000 different civilizations which are either trying to get in touch with us or waiting for a sign from other planets.

The terms of the Green Bank Formula take into account all the aspects in question, but in addition the scientists allotted two values to each term: a normal value admissible according to our present state of knowledge and an absolute minimum value.

$$N = R_{+} f_p n_e f_l f_i f_c L$$

In this formula:

R_{+} = the average annual number of new stars that are like our sun;

f_p = the number of stars with possible living beings;

n_e = the average number of planets which orbit the ecosphere of their sun and so have adequate premises for the development of life by human standards;

f_l = the number of planets favored in this way on which life has actually developed;

f_i = the number of planets which are populated by intelligences with their own ability to act during the lifetime of their sun;

f_c = the number of planets inhabited by intelligences that already have a developed technical civilization;

L = the life-span of a civilization, for only very long-lasting civilizations could encounter each other, given the vast distances in the universe.

If we take the lowest possible figures for all terms in this formula, we get: $N = 40$.

But if we take the admissible maximum value, we get: $N = 50,000,000$.

In other words, in the most unfavorable case the fantastic Green Bank Formula calculates that there are forty groups of intelligences in our Milky Way who are seeking contact with other intelligences.

The most audacious possibility gives 50,000,000 unknown intelligences who are waiting for a sign from the cosmos. All the Green Bank calculations are based not on present astronomical figures but on the number of stars in our Milky Way since it existed.

If we accept the formula of this scientific brain trust, civilizations with more advanced technologies than ours may have existed hundreds of thousands of years ago—a fact that supports the theory put forward here of visits by "gods" from the cosmos in the dawn of time. American astrobiologist Dr. Carl Sagan assures us that according to statistical calculations alone the possibility exists that our earth may have been visited by representatives of an extraterrestrial civilization at least once in the course of its history. Fantasy and wishful thinking may be concealed in all the deliberations and suppositions, but the Green Bank Formula is a mathematical formula and thus far removed from mere speculation.

A new branch of science is in the process of formation—what is known as exobiology. New branches of science always find it difficult to achieve recognition. Exobiology would certainly find it harder to find acceptance if recognized personalities were not already devoting their work to this new field of research which tackles extraterrestrial life with complete impartiality. What better proof of the seriousness of this new science than a group of the names which subscribe to it:

Dr. Freeman Quimby (chief of the NASA exobiological program), Ira Blei (NASA), Joshua Lederberg (NASA), L. P. Smith (NASA), R. E. Kaj (NASA), Richard Young (NASA), H. S. Brown (California Institute of Technology), Edward Purcell (professor of physics at Harvard University), R. N. Bracewells (Radio Astronomy Institute of Stanford), Charles Townes (Nobel Prize for physics, 1964), I. S. Shklovskii (Sternberg Institute, Moscow), N. S. Kardashev (Sternberg Institute, Moscow), Sir Bernard Lovell (Jodrell Bank), Wernher von Braun, Hermann Oberth, Von Braun's teacher, Ernst Stuhlinger, Eugene Sänger, and many others.

These names are representative of many thousands of exobiologists all over the world. The desire of all these men is to break through the taboos, to tear down the walls of lethargy which until now have always surrounded the desert areas of research which are specifically singled out in this book. In the face of all the opposition, exobiology exists, and one day it may become the most interesting and important field of research.

But how can a proof of life in the universe be produced until someone has been there? There are statistics and calculations that definitely favor the idea of extraterrestrial life. There is the evidence of bacteria and spores in space. The search for unknown intelligences has begun but has not yet produced results that are measurable, demonstrable, and convincing. What we need are verifications of theories—proofs of suppositions still disqualified as Utopian today. NASA has a research program ready that is intended to produce evidence of unknown life in the cosmos. Eight different probes, each one as unique as it is complicated, are to show evidence of life on planets in our solar system.

The following are the probes planned:

Optical Rotary Dispersion Profiles
The Multivator
The Vidicon Microscope
The J-Band Life Detector
The Radioisotope Biochemical Probe
The Mass Spectrometer
The Wolf Trap
The Ultraviolet Spectrophotometer

A few hints as to what is hidden behind these technical titles that are double Dutch to the layman:

Optical Rotary Dispersion Profiles is the name of a laboratory probe with a rotary searchlight. Once landed on a planet, this light begins to emit beams and search for molecules. Molecules are well-known prerequisites for every kind of life. One of these molecules is the large spiral-shaped DNS molecule, which consists of three chemical combinations arranged next to one another: nitrogen, sugar, and phosphoric acid. When the radiated light strikes such a molecule, the search beam is interrupted, because the nitrogen base adenin in chemical association with sugar has an "optically active" effect. Since the sugar combination in the DNS molecule is "optically active," the search beam of the probe has only to encounter a sugar-adenin combination to produce an immediate signal that, automatically sent to earth, would provide proof of life on an unknown planet.

The Multivator consists of a probe weighing barely 1 pound which is carried by a rocket as light baggage and ejected when near the planet. This miniature laboratory is then in a position to conduct as many as fifteen different experiments and transmit their results to earth.

The Radioisotope Biochemical Probe is the official name of a probe developed under the nickname Gulliver. It is intended to carry out a soft landing on the surface of another planet and immediately afterward to shoot out three 45-foot-long sticky ropes in various directions. In a few minutes these ropes will be automatically withdrawn into the probe; whatever stays clinging to the ropes—dust, microbes, or any kind of biochemical substances—will be immersed in a liquid culture medium. A part of this culture solution is enriched with the radioactive carbon isotope C-14; the microorganisms introduced would logically have to produce carbon dioxide, Co_2, through their metabolism. The gas carbon dioxide can easily be separated from the liquid culture and led to a measuring apparatus which measures the radioactivity of the gas containing C-14 nuclei and radios the results to earth.

The Wolf Trap, a mini-laboratory, was originally called Bug Detector by its inventor, but his collaborators rechristened it because their chief is called Professor Wolf Vishniac. The Wolf Trap is also supposed to make a soft landing on another planet and then extend a vacuum tube with a very fragile tip. When the tube touches the ground, the tip breaks and soil samples of all kinds will be sucked into the vacuum created. Once again the probe contains various sterile culture mediums which guarantee every kind of bacteria a rapid growth. This multiplication of the bacteria makes the liquid medium cloudy, and the pH value (the degree of acidity) of the liquid also changes. Both changes can be easily and accurately measured—the cloudiness of the liquid with the help of a beam of light and a photocell, the change in the acid content by an electrical pH measurement. These results would also enable us to make conclusions about existing unknown life.

Millions of dollars will be spent on the NASA program and coordinated research for the investigation and proof of extraterrestrial life. The first bioprobes will be sent to Mars. Undoubtedly man will soon follow the mini-laboratories which are the forerunners. The senior officials of NASA are unanimous in saying that the first astronauts will land on Mars by September 23, 1986, at the latest. This precise date has a reason. 1986 will be a year with little solar activity. Dr. Von Braun supports the view that men could land on Mars as early as 1982; NASA does not lack the technical resources, only an adequate and unbroken financial grant from Congress. In addition to all the current American responsibilities, two money swallowers such as the war in Vietnam and the space program are a heavy burden for even the richest nation in the world.

The plan for travel to Mars exists. The Mars spaceship has been designed. It "only" needs to be built as well. A model of it stands on the desk of an unusual man in Huntsville—Dr. Ernst Stuhlinger. Stuhlinger is director of the Research Project Laboratory, which is part of the George C. Marshall Space Flight Center in Huntsville, Alabama. He employs more than 100 scientific collaborators in his laboratories. They experiment in plasma-, nuclear-, and thermophysics and also occupy themselves with the basic research for projects pointing into the future. The research for the electric rocket motor of the future is forever linked with the name of Dr. Stuhlinger. He is the designer of the Mars spaceship which will carry men to the red planet in our century.

Dr. Stuhlinger was brought to the United States soon after World War II by his friend Dr. Wernher von Braun; in Fort Bliss they made rockets for the Air Force. Accompanied by 162 fellow countrymen, the two rocket pioneers moved to Huntsville after the outbreak of the Korean War in order to conjure up a project such as even America, accustomed as it is to gigantomania, has never seen before.

In those days Huntsville was a small, sleepy nest on the edge of the Appalachian Mountains. With the arrival of the rocket men the little cotton town turned into a circus. Factories, rocket testing platforms, laboratories, giant hangars, and corrugated iron offices shot up from the ground breathtakingly fast in a few years. Today more than 150,000 people live in Huntsville; the little town has awakened from its sleep and its inhabitants have become enthusiastic space fans. When the first Redstone rockets thundered away from the testing platform, many Huntsvillites ran down into their cellars in panic. Nowadays, when a Saturn rocket is tested and a roar fills the air as if the world were coming to an end in the next second, nobody takes any notice. Huntsvillites always carry their earplugs with them, just as Londoners carry their umbrellas. They call their town simply Rocket City, and if Congress does not grant the hundreds of thousands of millions of dollars demanded for space travel, they get bad-tempered and start agitating. They have every reason to be proud of their "Germans" and NASA, for Huntsville has grown into the biggest NASA center of all. Here the rockets that make headlines all over the world are thought up and designed, from the little Redstone to the gigantic Saturn V. At launching, the tanks are filled with some 880,000 gallons of highly explosive fuel, which develops a propulsive force of 150,000,000 horsepower. The giant rocket weighs almost 3,000 tons. In Huntsville some 7,000 technicians, engineers, and scientists of related disciplines are working under Wernher von Braun toward the great goal, the conquest of space. In 1967 about 300,000 scientists of all kinds were working on America's global space program. More than

20,000 industrial firms are working for the greatest research undertaking in history.

When I visited Huntsville, Austrian scientist Dr. Pscherra told me that the research groups constantly had to develop new "articles" which had never before been produced anywhere in the world.

"Look here!" he said and showed a large cylinder from which came a humming, rumbling noise. "In there we are conducting lubricating experiments in an absolute vacuum. Do you know that we cannot use any of the countless lubricants produced in the world? They lose all their lubricating qualities in space. With available lubricants even a simple electromotor stops functioning after at most half an hour in airless space. What else could we do but invent a lubricant which works perfectly even in an absolute vacuum?"

A terrible groaning and whining came from another room. Two tremendous vises, firmly anchored to the floor, were trying to pull a four-inch-thick sheet of metal to pieces.

"Another series of experiments that we would willingly dispense with," said Dr. Pscherra, "but our experience has shown that existing metal alloys do not stand up to the stresses of space. So we must find ones that meet our requirements. That is the reason for these tensile probes and fatigue experiments under every conceivable kind of space situation. We also have to develop new welding techniques. The welded joints must be subjected to cold, heat, vibration, tensile-strength and pressure tests, so that we can find out the limits at which a welded joint breaks."

The hostess who accompanied me looked at her watch. Dr. Pscherra looked at his watch. Everyone was looking at his watch. NASA personnel, of course, don't notice it anymore; the visitor finds it curious at first, but he soon gets used to it, for looking at one's watch is a standard gesture of NASA personnel at Cape Kennedy, Houston, and Huntsville. They always seem to be making a countdown: four . . . three . . . two . . . one . . . zero.

Rides and walks through halls, corridors, and doors led, after many more security controls, to Mr. Pauli, who also comes from German-speaking Europe and has been working for NASA for thirteen years. I had a white helmet bearing the NASA symbol crammed on my head; Mr. Pauli took me to the testing platform of the Saturn V. The simple words "testing platform" mean a concrete colossus that weighs several hundred tons, is several stories high, has lifts and cranes leading to it, and is surrounded by ramps in which a bewildering network of many miles of cables is installed. Once it is ignited, Saturn V makes a din which can be heard 12 miles from the launching ground. The testing platform, deeply anchored in rock and concrete, rises as much as three inches from its base during such trials, while 333,000 gallons of water per second are pumped through a sluice for cooling purposes. Merely for cooling trial rockets on the testing platform, NASA had to build a pumping works that could supply a city the size of Manchester with drinking water. A single firing test costs a cool $14,000,000; space does not come cheaply.

Huntsville is one of the many NASA centers. The reader should note them because later they may become the departure stations for space flights:

Army Research Center, Moffet Field, California
Electronics Research Center, Cambridge, Massachusetts
Flight Research Center, Edwards, California
Goddard Space Flight Center, Greenbelt, Maryland
Propulsion Laboratory, Pasadena, California
John F. Kennedy Space Center, Florida
Langley Research Center, Hampton, Virginia
Lewis Research Center, Cleveland, Ohio
Manned Spacecraft Center, Houston, Texas

Nuclear Rocket Development Station, Jackass Flats, Nevada

Pacific Launch Operations Office, Lompoc, California

Wallops Station, Wallops Island, Virginia

Western Operations Office, Santa Monica, California

NASA Headquarters, Washington, D.C.

The spaceship industry has long overtaken the automobile industry as a pacesetter in the market. On July 1, 1967, 22,828 people were working at the Cape Kennedy Space Center; the annual budget for this station alone amounted to $475,784,000 in 1967!

All that because a few crazy people want to go to the moon? I think I have already given sufficient convincing examples of what we owe research into space travel today (and these only as by-products), ranging from objects in everyday use to complicated medical apparatus which saves people's lives every hour of the day all over the world. The supertechnology in the course of development is truly no scourge of mankind. It is carrying mankind into the future which begins anew daily with seven-league boots.

The author had a chance to ask Wernher von Braun for his opinion of the hypotheses put forward in this book:

"Dr. Von Braun, do you consider it possible that we shall find life on other planets in our solar system?"

"I consider it possible that we shall come across lower forms of life on the planet Mars."

"Do you consider it possible that we are not the only intelligences in the universe?"

"I consider it extremely probable that not only plant and animal life but also intelligent living creatures exist in the infinite reaches of the universe. The discovery of such life is a most fascinating and interesting task, but considering the enormous distances between our own and other solar systems and the still greater distances between our galaxy and other galactic systems, it is doubtful whether we shall succeed in proving the existence of such forms of life or getting into direct communication with them."

"Is it conceivable that older, technically more advanced intelligences live or have lived in our galaxy?"

"Up to the present we have no proof or indication that older and technically more advanced living beings than ourselves live or have lived in our galaxy. However, on the basis of statistical and philosophical considerations, I am convinced of the existence of such advanced living beings. But I must emphasize that we have no firm scientific basis for this conviction."

"Is there a possibility that older intelligences could have paid a visit to our earth in the dim mists of time?"

"I won't deny this possibility. But to the best of my knowledge no archaeological studies have so far provided any basis for that kind of speculation."

Here my conversation with the "father of the Saturn rocket" ended. I could not tell him in detail about all the remarkable discoveries, the absurdities, the old books handed down to us as unsolved puzzles—the countless questions that archaeological finds force upon us when considered with space eyes. But Dr. Von Braun awaits the documentation of this book.

12

Tomorrow

WHERE do we stand today?

Will man dominate space one day?

Did unknown beings from the infinite reaches of the cosmos visit the earth in the remote past?

Are unknown intelligences somewhere in the universe trying to make contact with us?

Is our age, with its discoveries that are taking the future by storm, really so terrible?

Should the most shattering results of research be kept secret?

Will medicine and biology discover a way of restoring deep-frozen men to life?

Will men from earth colonize new planets?

Will they mate with the inhabitants they find there?

Will men create a second, third, and fourth earth?

Will special robots replace surgeons one day?

Will hospitals in the year 2100 be spare-part stores for defective men?

Will it become possible in the distant future to prolong

man's life indefinitely with artificial hearts, lungs, kidneys, etc.?

Will Huxley's *Brave New World* come true one day in all its improbability and chilling inhumanity?

A compendium of such questions could easily get as big as the New York telephone directory. Not a day passes without something brand new being invented somewhere in the world—every day another question can be struck from the list of impossibilities as answered. Edinburgh University received a preliminary grant of $6,480,000 from the Nuffield Trust for the development of an intelligent computer. The prototype of this computer was put into conversation with a patient, and afterward the patient would not believe that he had been talking to a machine. Dr. Donald Michie, who designed this computer, claimed that his machine was beginning to develop a personal life.

The new science is called futurology! It's goal is the planning and detailed investigation and understanding of the future by all the technical and mental means available. Think tanks are springing up all over the world; what they amount to are monasteries of scientists of today, who are thinking for tomorrow. There are 164 of these think tanks at work in America alone. They accept commissions from the government and heavy industry. The most celebrated think tank is the Rand Corporation at Santa Monica in California. The U.S. Air Force was responsible for its foundation in 1945. The reason? High-ranking officers wanted a research program of their own on intercontinental warfare. Some 850 selected scientific authorities now work in the two-story magnificently laid-out research center. The first ideas and plans for the foundations of mankind's most improbable adventures are born here. As early as 1946 Rand scientists evaluated the military usefulness of a spaceship. When Rand developed the program for various satellites in 1951, it was called Utopian. Since Rand has been functioning, the world can thank this research center for 3,000 accurate accounts of hitherto unobserved phenomena. Rand scientists have published more than 110 books, which have advanced our culture and civilization immeasurably.

There is no end in sight to this research work, and there is unlikely to be one.

Similar work for the future is being done in the following institutes: The Hudson Institute at Harmon-on-Hudson, New York; the Tempo Center for Advanced Studies belonging to General Electric at Santa Barbara, California; the Arthur Little Institute at Cambridge, Massachusetts, and the Battelle Institute at Columbus, Ohio.

Governments and big business simply cannot manage without these thinkers for the future. Governments have to decide on their military plans far in advance; big businesses have to calculate their investments for decades ahead. Futurology will have to plan the development of capital cities for a hundred or more years ahead.

Equipped with present-day knowledge, it would not be difficult to estimate, say, the development of Mexico for the next fifty years. In making such a forecast, every conceivable fact would be taken into account, such as the existing technology, means of communication and transport, political currents, and Mexico's potential opponents. If this forecast is possible today, unknown intelligences could certainly have made such a forecast for the planet Earth 10,000 years ago.

Mankind has a compulsive urge to think out in advance and investigate the future with all the potentialities at its command. Without this study of the future, we would probably have no chance of unraveling our past. For who knows whether important clues for the unraveling of our past do not lie around the archaeological sites, whether we do not trample them heedlessly under foot because we do not know what to make of them.

That is the very reason why I advocated a "Utopian archaeological year." In the same way that I am unable to "believe" in the wisdom of the old patterns of thought, I do not ask others to "believe" my hypothesis. Nevertheless, I expect and hope that the time will soon be ripe to attack the riddle of the past without prejudice—making full use of all the refinements of technology.

It is not our fault that there are millions of other planets in the universe.

It is not our fault that the Japanese statue of Tokomai, which is many thousands of years old, has modern fastenings and eye apertures on its helmet.

It is not our fault that the stone relief from Palenque exists.

It is not our fault that Admiral Piri Reis did not burn his ancient maps.

It is not our fault that the old books and traditions of human history exhibit so many absurdities.

But is our fault if we know all this but disregard it and refuse to take it seriously.

Man has a magnificent future ahead of him, a future which will far surpass his magnificent past. We need space research and research into the future and the courage to tackle projects that now seem impossible. For example, the project of concerted research into the past which can bring us valuable memories of the future. Memories which will then be proved and which will illuminate the history of mankind—for the blessing of future generations.

Bibliography

ALLEN, T., *The Quest,* Philadelphia, Chilton Books, 1965.

BACON, E., *Auferstandene Geschichte,* Orell Füssli, 1964.

BASS, G. F., *Archaeology under Water,* Thames & Hudson, 1966.

BELLAMY, H. S., and ALLAN, P., *The Great Idol of Tiahuanaco,* Faber & Faber, 1959.

BETZ, O., *Offenbarung und Schriftforschung der Qumransekte,* Mohr, 1960.

BOSCHKE, F. L., *Erde van anderen Sternen,* Econ, 1965.

BRAUN, WERNHER VON, *The Next 20 Years of Interplanetary Exploration,* Marshall Space Flight Center, Huntsville, 1965.

BURROWS, M., *Mehr Klarheit über die Schriftrollen,* Beck, 1958.

CHARROUX, ROBERT, *Histoire inconnue des hommes depuis cent mille ans,* Laffont, Paris.

———, *Le livre des secrets trahis,* Robert Laffont, Paris, 1965.

CHARDIN, TEILHARD DE, *Phenomenon of Man,* Collins, 1961.

CLARK, GR., "Die ersten 500,000 Jahre," from: *Die Welt aus der wir kommen,* Knaur, 1961.

CLARKE, ARTHUR C., *The Challenge of the Space Ship,* Harper & Brothers, Publishers, New York.

———, *Profiles of the Future,* Victor Gollancz Ltd., London, 1962.

———, *Voices from the Sky,* Harper & Row, Publishers, New

York.

CORDAN, W., *Das Buch des Rates, Mythos und Geschichte der Maya,* Diederichs, 1962.

COTTRELL, L., *The Anvil of Civilisation,* New English Library, 1967.

CYRIL, A., "Gott-Könige besteigen den Thron," from: *Die Welt aus der wir kommen,* Knaur, 1961.

DUPONT, A., *Les écrits esseniens découverts près de la mer morte,* Payot, 1959.

DUTT, M. NATH, *Ramayana,* Calcutta, 1891.

EINSTEIN, A., *Grundzüge der Relativitätstheorie,* Vieweg, 1963.

FALLACI, O., *Wenn die Sonne stirbt,* Econ, 1966.

HAPGOOD, C. H., *Maps of the Ancient Sea Kings,* New York, Chilton Books, 1965.

HEINDEL, M., *Die Weltanschauung der Rosenkreuzer,* Zurich, Rosenkreuzer.

HERODOTUS, *Historien,* Books I-IX.

HERTEL, J., *Indische Märchen,* Diederichs, 1961.

HEYERDAHL, THOR, *Aku-Aku,* Allen & Unwin, 1958.

HOENN, K., *Sumerische und akkadische Hymnen und Gebete,* Artemis, 1953.

KELLER, W., *The Bible as History,* Hodder & Stoughton, 1956.

KÜHN, H., *Wenn Steine reden,* Brockhaus, 1966.

LEY, WILLY, *Die Himmelskunde,* Econ, 1965.

LHOTE, H., *Die Felsbilder der Sahara,* Zettner, 1958.

LOHSE, E., *Die Texte aus Qumran,* Kösel, 1964.

LOVELL, SIR B., *The Exploration of Outer Space,* Oxford University Press, 1962.

MALLOWAN, M. E. L., "Geburt der Schrift, Geburt der Geschichte," from: *Die Welt aus der wir kommen,* Knaur, 1961.

MASON, J. A., *The Ancient Civilizations of Peru,* Penguin Books, 1957.

MELLAART, J., "Der Mensch schlägt Wurzel," from: *Die Welt aus der wir kommen,* Knaur, 1961.

PAKRADUNY, T., *Die Welt der geheimen Mächte,* Tiroler Graphik, 1952.

PAUWELS, L., and BERGIER, J., *Aufbruch ins dritte Jahrtausend,* Scherz, 1962.

REICHE, M., *The Mysterious Markings of Nazca,* New York, 1947.
———, *Mystery on the Desert,* Lima, 1949.
ROY, P. CH., *Mahabharata,* Calcutta, 1889.
RÜEGG, W., *Die ägyptische Götterwelt,* Artemis, 1959.
———, *Kulte und Orakel im alten Ägypten,* Artemis, 1960.
SÄNGER, E., *Raumfahrt, heute, morgen, übermorgen,* Econ, 1963.
SANTA DELLA, E., *Viracocha,* Brussels, 1963.
SCHENK, G., *Die Grundlagen des XXI. Jahrhunderts,* Deutsche Buchgemeinschaft, 1965.
SCHWARZ, G. TH., *Archäologen an der Arbeit,* Franke, 1965.
SHAPLEY, H., *Wir Kinder der Milchstrasse,* Econ, 1965.
SHKLOVSKII, I. S., and SAGAN, C., *Intelligent Life in the Universe,* San Francisco, 1966.
SUGRUE, TH., *Edgar Cayce,* Dell, 1957.
TOZZER, A. M., *Chichén Itzá and Its Cenote of Sacrifice,* Memoirs of the Peabody Museum, Cambridge, Mass., 1957.
VELIKOVSKY, I., *Worlds in Collision,* Victor Gollancz, 1950.
WATSON, W., "Im Bannkreis von Cathay," from: *Die Welt aus der wir kommen,* Knaur, 1961.
WAUCHOPE, R., *Implications of Radiocarbon Dates, from Middle and South America,* Tulane University, New Orleans, 1954.
ZIEGEL, F. Y., *Nuclear Explosion over the Taiga,* U.S. Dept of Commerce, Office of Technical Services, 1962.

GENERAL READING

Kunst der Welt, 4 vols., Neue Schw. Bibli., 1960 and 1961.
Die Edda, 2 vols., Altnordische Dichtungen, Thule.
Gilgamesch, Epos der alten Welt, Insel.
Relación de las cosas de Yucatán, D. Landa, Mexico, 1938.
Das Mahabharata, Roy Biren, Diederichs, 1961.
Shells and Other Marine Material from Tikal, University of Pennsylvania, 1963.
Cichen Itza, Instituto Nacional de Antropologia e Historia Mexico, 1965.
Bhagavadgita, Diederichs, 1922.
Die Heilige Schrift des Alten Testamentes, Zwingli Zurich.

Traktatüber die Kriegskunst, Minst. für Nationale Verteidigung, Berlin, 1957.

Gulliver's Travels, Jonathan Swift, 1727.

Dokumentarbericht über den 4 internationalen UFO/IFO Kongress in Wiesbaden, Karl Veit, 1960.

Report from Mars, Mariner 1964–1965, Jet Propulsion Laboratory, California Institute of Technology.

The Search for Extraterrestrial Life, NASA, Washington, D.C.

NASA at the John F. Kennedy Space Center.

Meteorological Flight 1966 (NASA) Sounding Rockets, NASA, 1966.

Astronaut Training, Manned Spacecraft Center, Houston, Fact Sheet 290.

NASA—FACTS, Vol. II, No. 5; No. 13.

NASA Marshall Space Flight Center, Public Affairs Office, March 29, 1966.

NASA Marshall Space Flight Center, Public Affairs Office, Aug, 5, 1966.

NASA Marshall Space Flight Center, Public Affairs Office, Sept. 26, 1966.

NASA Marshall Space Flight Center, Public Affairs Office, Sept. 29, 1966.

Der Spiegel, No. 46, Nov. 6, 1967.

Stern, No. 47, Nov. 9, 1967.

Die Zeit, No. 46, Nov. 7, 1967; No. 47, Nov. 25, 1967; No. 51, Dec. 22, 1967.

Zweites Deutsches Fernsehen (German TV2): *"Invasion aus dem Kosmos?"* Nov. 6, 1967.

Suddeutsche Zeitung, Munich, Nov. 21/23, 1967.

Index